Genetic Prospects

Institute for Philosophy and Public Policy Studies
General Editor: Verna V. Gehring

About the Series. This new series grows out of a collaboration between the Institute for Philosophy and Public Policy at the University of Maryland and Rowman & Littlefield Publishers. Each slim volume in the series offers an insightful, accessible collection of essays on a current topic of real public concern, and which lies at the intersection of philosophy and public policy. As such, these books are ideal resources for students and lay readers, while at the same time making a distinctive contribution to the broader scholarly discourse.

About the Institute. Established in 1976 at the University of Maryland and now part of the School of Public Affairs, the Institute for Philosophy and Public Policy was founded to conduct research into the conceptual and normative questions underlying public policy formation. This research is conducted cooperatively by philosophers, policy makers and analysts, and other experts both within and outside of government. The Institute publishes the journal *Philosophy & Public Policy Quarterly* and the series Institute for Philosophy and Public Policy Studies with Rowman & Littlefield Publishers.

War after September 11

Genetic Prospects: Essays on Biotechnology, Ethics, and Public Policy

Genetic Prospects

Essays on Biotechnology, Ethics, and Public Policy

EDITED BY VERNA V. GEHRING

ROWMAN & LITTLEFIELD PUBLISHERS, INC.
Lanham • Boulder • New York • Toronto • Oxford

ROWMAN & LITTLEFIELD PUBLISHERS, INC.

Published in the United States of America
by Rowman & Littlefield Publishers, Inc.
A wholly owned subsidiary of The Rowman & Littlefield Publishing Group, Inc.
4501 Forbes Boulevard, Suite 200, Lanham, Maryland 20706
www.rowmanlittlefield.com

PO Box 317
Oxford
OX2 9RU, UK

British Library Cataloguing in Publication Information Available

Library of Congress Cataloging-in-Publication Data

Genetic prospects : essays on biotechnology, ethics, and public policy / edited by Verna V. Gehring.
p. ; cm.
Includes bibliographical references and index.
ISBN 0-7425-3334-4 (cloth : alk. paper) — ISBN 0-7425-3335-2 (paper : alk. paper)
1. Genetic engineering—Moral and ethical aspects. [DNLM: 1. Genetic Engineering—ethics. 2. Biotechnology—ethics. QH 442 G32952 2003] I. Gehring, Verna V.

QH442.G4596 2003
306.4'6—dc21 2003009740

Printed in the United States of America

™ The paper used in this publication meets the minimum requirements of American National Standard for Information Sciences—Permanence of Paper for Printed Library Materials, ANSI/NISO Z39.48-1992.

Contents

III: The Ethics (and Politics) of Genetic Technologies

Preface

Several of the essays in this work appeared as articles in *Philosophy & Public Policy Quarterly,* the journal sponsored by the Institute for Philosophy and Public Policy at the School of Public Affairs, University of Maryland. William A. Galston, director of the Institute for Philosophy and Public Policy, contributed the introduction, and research scholars of the Institute contributed four articles. This volume has benefited from the conversation and thought of all of the research scholars who, during informal paper talks, offered suggestions for improvement. Thanks must go to Robert Wachbroit for suggesting the title of the work, and to Carroll Linkins and Richard Chapman for their help and kindness in bringing the essays to press.

The Institute for Philosophy and Public Policy gratefully acknowledges the National Human Genome Research Institute program on Ethical, Legal, and Social Implications of Human Genetics, Grant R01HG02363, for its generous financial support for research that produced this volume. Although this project would not have been possible without NIH-ELSI support, the views expressed are those of the authors and not necessarily of any funding agency.

The editor especially thanks William A. Galston, Mark Sagoff, Robert Wachbroit, and David A. Wasserman for their helpful comments and thoughtful advice in the development of the volume.

Verna V. Gehring
Editor
Institute for Philosophy and Public Policy
School of Public Affairs, University of Maryland
College Park, Maryland

Introduction

William A. Galston

While the collapse of the Soviet Union has diminished the force of George Orwell's *1984,* the other great dystopian tract of the twentieth century, Aldous Huxley's *Brave New World,* is timelier than ever. The ongoing progress of genetic science may well revolutionize medicine and human reproduction, and it may end by giving us the ability to transform the human species itself.

This new power has raised hopes that we will solve a range of genetically based problems that afflict us. It has also evoked fears that we are on the verge of a "posthuman" future in which precious but precarious norms regulating individual and social life will be set aside. Some worry that the posthuman age will be hubristic, even Promethean, reveling in the unfettered self-creative liberty. Others fear just the reverse, that it will lower our aspirations by tempting us to place comfort, health, and longevity above more exalted but less self-regarding aims. The spirit of both Nietzsche's "overman" and his "last man" hovers over this discussion.

The question is whether we will have the moral and political wisdom to avoid these two extremes in our use of new biotechnologies. This question depends, in turn, on our ability to think clearly about the fundamental issues involved and to draw tolerably clear and defensible lines between what is to be permitted and what prohibited.

What are the resources from which the needed norms and maxims might be drawn? One possibility: from common sense. Prudential considerations based on uncertainty about consequences and about the balance of costs and benefits from genetic interventions should be given their due. If experience has taught us caution about intervening in the natural environment, then it would seem all the more necessary to be cautious when the arena of intervention is our own species. These considerations have led thinkers such as Francis Fukuyama to recommend a precautionary principle: Genetic modification of human beings should be presumed guilty until proven innocent, and those who propose new interventions must accept and discharge the burden of showing that their promise outweighs their peril.

Prudential arguments suffer from some obvious shortcomings. They may yield at most imprecise and shifting norms, and they cannot address our deepest fears. These considerations have led some thinkers to seek guidance from religion—especially in the United States, by far the most religious of the Western democracies. But our religious diversity matches our religious fervor, and not all religious traditions agree about the extent to which biotechnology should be controlled. While Catholicism and fundamentalist Protestantism share restrictive views on this matter, Judaism (including Jewish Orthodoxy, which makes common cause with Catholics and conservative evangelicals on many other issues) is much more permissive. Judaism accords a very high value to human reproduction and healing, and a relatively low moral status to nonimplanted pre-embryos and to fetuses in an early stage of development. It also sets forth an activist vision of human beings as co-creators of the world. These basic features of the Jewish outlook lead it to embrace many practical applications of biotechnology. Beyond Judaism, there are many mainstream and liberal Protestant denominations that do not accept the stance of the Catholic-conservative evangelical entente. I have not yet mentioned the demographic growth of Islam, Hinduism, and Buddhism in American society, which will further diversify our public discourse over the next generation. There is, then, no unitary or canonical religious teaching on which we can base the public regulation of genetic technology.

Third, we might seek guidance from what the eighteenth-century philosopher Adam Smith and others have termed "moral sentiments." It might be the case that human beings share a range of responses to morally fraught situations and that these responses form the core of a common moral code. In this vein, Leon Kass, now chairman of

President George W. Bush's Council on Bioethics, writes of the "wisdom of repugnance" and suggests that this instinctive moral reaction serves to limit the excesses of genetic technology.

One difficulty with this proposal is that not all human beings feel repugnance about the same things, or with similar intensity. A deeper difficulty is that even when a community does share repugnance over a specific practice, it may nonetheless be mistaken. Two cases are instructive. There was a time, not so long ago, when entire communities in the United States viscerally opposed racially mixed marriages. We can now see that the policies encoding these sentiments were wrongheaded and destructive. Centuries ago, the dissection of corpses for medical purposes was considered deeply repugnant. While this sentiment does not seem as wrong as feeling disgust over mixed marriages, we can still reject the policy implications that many communities drew from it and enforced. Even when repugnance suggests the presence of something genuinely troubling, it may not be dispositive in practice. Our moral sentiments are at most the beginning of wisdom, and they call for the kind of critical reflection that might isolate the generally defensible principles to which they point.

This brings us to the fourth possible resource from which we might draw insight into genetic technology—namely, secular philosophy. While hardly devoid of respect for (or references to) religion, the essays in this volume lie squarely in the philosophical camp. Broadly speaking, they address three kinds of questions: the implications of genetic science for our understanding of nature; its impact on our conception of human nature; and the challenges it poses for specific issues of private conduct or public policy.

Nature. Taking as his point of departure the public controversy over genetically modified food, Mark Sagoff distinguishes four different conceptions of nature: everything in the universe (as distinct from the "supernatural"); creation in the sense of what God has made—the sacred as opposed to the profane; that which is independent of human influence or contrivance—pristine as opposed to spoiled, natural as opposed to artificial; and finally, that which is authentic or true to itself, as opposed to the specious and superficial. Sagoff points out that these four conceptions are not only logically independent of one another in theory but point in different directions in practice. Adverting to the first of Sagoff's four conceptions, the makers of genetically modified food insist that their products are natural, while those who focus on nature as the opposite of human contrivance vehemently disagree.

Like Sagoff, Paul Thompson locates his reflections in the agricultural sphere. While broadly accepting Sagoff's fourfold way, Thompson offers a fifth conception of nature, drawn from farming. He offers an analogy between farmers and other artisans who work with natural materials. For such artisans, "To work against nature is to go against the grain." For example, "wood and stone materials have a natural grain that determines physical properties such as tensile or compression strength." To ignore these properties is to risk creating objects that will not last and are ill suited to the purposes for which they were made. Similarly, farmers must work with the grain of their chosen materials, which will require a substantial measure of inductive local knowledge as opposed to universal scientific propositions. It is not clear, Thompson emphasizes, whether this conception of nature as rooted in artisan knowledge is equally applicable to areas such as medicine, where the balance between general and local knowledge, science and practical experience, may prove quite different.

Human nature. Harold Baillie argues that policies addressing genetic engineering "must reflect some vision of what human beings are." He suggests that the notion of the "sacred" is an important part of this inquiry because it identifies "that element in . . . human nature that ought to be preserved." After working through, and discarding, a number of conceptions of the sacred that he regards as unsustainable, Baillie focuses on the human interest in transcendence—a deep psychological force, akin to Platonic *eros,* that drives human beings to go beyond the ordinary. As desiring beings, we realize that nature, particularly our physical nature, is at times "overwhelming in its capriciousness," and we struggle against its arbitrary harms. If we are to generate principles and practices that allow us reasonably to govern the use of genetic technologies, Baillie concludes, we must find ways to address "both our striving for beauty and immortality, and our revulsion at a nature indifferent beyond measure."

Robert Wachbroit agrees with Baillie that the issues of genetic technology and human nature are intimately linked. Not only might we hope to derive from human nature norms for the regulation of genetic technology, but also discoveries in this area can affect our self-conception. Wachbroit distinguishes multiple meanings—statistical, social, and biological—of the "normal"—and argues against straightforwardly equating statistical normality with nature. And while advances in human genetics can help us better understand human biological normality, they cannot determine the content of human nature, understood as "what is essential to being human." And even if such properties were located (say, through sophisticated interspecies genetic comparisons), their discovery would not directly entail social or moral conclusions. The relationship between human nature and public policy is complex.

Public policy. It remains to note, all too briefly, the essays that address specific public policy issues raised by advances in genetic science. After reviewing the principal arguments for and against human cloning, Richard Zaner concludes that only a comprehensive new ethics covering the environment, politics, and the market economy would have any serious chance of moving, in a more democratic direction, decisions otherwise made by small groups in corporate boardrooms and the halls of government.

Robert Wachbroit offers a similarly comprehensive look at the controversy over stem cell research. While acknowledging some overlap, he resists the fashionable but, he believes, misleading analogy between this issue and the abortion controversy. Societal acceptance of *in vitro* fertilization suggests that the moral status of the preimplantation embryo is not the crucial question in assessing stem cell research. In practice, the central issue may well be "complicity"—the ways in which we are willing (or unwilling) to derive advantages from practices many citizens regard as morally questionable.

Deborah Hellman poses a question that progress in understanding the genetic basis of disease will render increasingly urgent over the next decade: What (if anything) is especially wrong about genetic discrimination? In particular, if it is acceptable to distinguish between sick and healthy individuals for purposes of health insurance, why is it wrong to distinguish between those with and without genetic predispositions to the same illnesses? After working through, and rejecting, a wide range of familiar answers to this question, Hellman suggests that the distinctive wrongfulness of genetic discrimination lies in the association it expresses with the tangled history of eugenics. Whatever the intention of the practice, its social meaning may well be that individuals with genetic flaws are less worthy than others. And that is unacceptable.

David Wasserman poses another question, the urgency of which is bound to grow over time: What is wrong with parents genetically enhancing their children? Like Hellman, he contends that several familiar arguments are unpersuasive. To the extent that the practice of genetically enhancing children is morally questionable, he suggests, it is because it represents an excess of control. In genetics as in child rearing, parents should "strive for a means between control and acquiescence." Wasserman's Aristotelian language suggests, not moral bright lines, but rather the exercise of judgment, which is bound to prove contestable.

Finally, Sara Goering wrestles with a question that few are yet asking but which citizens will surely face in the future: What norms should control the use of genetic technology for aesthetic or cosmetic purposes? She suggests that the doubts we harbor about the moral merits of cosmetic surgery apply as well to the prospect of cosmetic genetics. Specifically: "if a practice contributes to or reinforces harmful conceptions of normality, we should look for other means to achieve individual interests." Goering's argument suggests that in addressing this question, the language of prudence and virtue may be just as relevant as the language of choices and prohibitions. Not just public regulation, but also public education, will be necessary if we are to achieve a human society in which individuals do not experience harmful conceptions of normality as socially coercive.

Some of the essays in this volume are self-consciously exploratory, yielding at most tentative and fragmentary results. While others adopt a decidedly firmer tone, they may evoke as much resistance as assent. Readers are likely to come away with more questions and objections than answers. If so, they may well ask, how has philosophical inquiry advanced us beyond the limits of common sense, the vagaries of moral sentiments, or the deep disagreements among religious traditions?

The response (such as it is) rests on a distinction between two conceptions of the public role of philosophy. One conception, which traces back to Plato, is that philosophy prescribes to the public realm. To do so, its voice must be clear, unequivocal, and undivided. The other conception, with a more varied ancestry, is that philosophy makes a distinctive but not authoritative contribution to public life by clarifying concepts, exploring links between assertions, and reducing avoidable misunderstandings. So understood, public philosophy is an important voice in the dialogue of democracy, and it can carry on its work without reaching agreement among philosophers or forcing agreement among citizens.

This is to suggest a fifth possible source of the norms we need to control genetic technologies—our best collective judgment, all things considered, as worked out and expressed through democratic institutions. If citizens in the aggregate are wiser than any individual citizen, as Aristotle suggests, then our democratic institutions—suitably informed by philosophy, religion, sentiment, and shared experience—represent our best (if imperfect) hope for creating a sustainable regime of genetic regulation.

The Concept of the Natural

I

Genetic Engineering and the Concept of the Natural

Mark Sagoff

Why do many consumers view genetically engineered foods with suspicion? I want to suggest that it is largely because the food industry has taught them to do so. Consumers learn from advertisements and labels that the foods they buy are all natural—even more natural than a baby's smile. "The emphasis in recent years," *Food Processing* magazine concludes, "has been on natural or nature-identical ingredients." According to *Food Product Design*, "the desire for an all natural label extends even to pet food."

The food industry, I shall argue, wishes to embrace the efficiencies offered by advances in genetic engineering. This technology, both in name and in concept, however, belies the image of nature or of the natural to which the food industry constantly and conspicuously appeals. It should be no surprise that consumers who believe genetically modified foods are not "natural" should for that reason regard them as risky or as undesirable. If they knew how much technology contributes to other foods they eat, they might be suspicious of them as well.

All-Natural Technology

Recently, I skimmed through issues of trade magazines, such as *Food Technology* and *Food Processing*, that serve the food industry. In full-page advertisements, manufacturers insist the ingredients they market come direct from primordial Creation or, at least, that their products are identical to nature's own. For example, Roche Food Colours runs in these trade magazines a full-page ad that displays a bright pink banana over the statement: "When nature changes her colours, so will we." The ad continues:

> Today more and more people are rejecting the idea of artificial colours being used in food and drink. . . .
>
> Our own food colours are, and always have been, strictly identical to those produced by nature.
>
> We make pure carotenoids which either singly or in combination achieve a whole host of different shades in the range of yellow though orange to red.
>
> And time and time again they produce appetising natural colours, reliably, economically, and safely.
>
> Just like nature herself.

Advertisement after advertisement presents the same message: food comes directly from nature or, at least, can be sold as if it did. Consider, for example, a full-page advertisement that McCormick and Wild, a flavor manufacturer, runs regularly in *Food Processing*. The words "BACK TO NATURE" appear under a kiwi fruit dripping with juice. "Today's consumer wants it all," the advertisement purrs, "great taste, natural ingredients, and new ideas. . . . Let us show you how we can put the world's most advanced technology in natural flavors at your disposal. . . . "

This advertisement clearly states the mantra of the food industry: "Today's consumer wants it all." Great taste. Natural ingredients. New ideas. The world's most advanced technology. One can prepare the chemical basis of a flavor, for example, benzaldehyde—almond—artificially with just a little chemical know-how, in this instance, by mixing oil of clove and amyl acetate. To get exactly the same compound as a "natural" flavor, one must employ far more sophisticated technology to extract and isolate benzaldehyde from peach and apricot pits. The "natural" flavor, an extract, contains traces of hydrogen cyanide, a deadly poison evolved by plants to protect their seeds from insects. Even so, consumers strongly prefer all-natural to artificial flavors, which sell therefore at a far lower price.

In its advertisements, the Haarmann & Reimer Corporation (H&R) describes its flavor enhancers as "HypR Clean Naturally." With "H&R as your partner, you'll discover the latest advances in flavor technology" that assure "the cleanest label possible." A "clean" label is one that includes only natural ingredients and no reference to technology. In a competing advertisement, Chr. Hansen's Laboratory announces itself as the pioneer in "cultural and enzyme technologies. . . . And because our flavors are completely natural, you can enjoy the benefits of 'all-natural' labeling." Flavor manufacturers tout their stealth technology—i.e., technology so advanced it disappears from the consumer's radar screen. The consumer can be told he or she is directly in touch with nature itself.

The world's largest flavor company, International Flavors & Fragrances (IFF), operates manufacturing facilities in places like Dayton, New Jersey, an industrial corridor of refineries and chemical plants. Under a picture of plowed, fertile soil, the IFF Laboratory, in a full-page display, states, "Where Nature is at work, IFF is at work." The text describes "IFF's natural flavor systems." The slogan follows: "IFF technology. In partnership with Nature." Likewise, MEER Corporation of Bergen, New Jersey, pictures a rainforest under the caption, "It's A Jungle Out There!" The ad states that "true-to-nature" flavorings "do not just happen. It takes . . . manufacturing and technical expertise and a national distribution network . . . for the creation of natural, clean label flavors."

Food colors are similarly sold as both all natural and high tech. "VegetoneH colors your foods *naturally* for a healthy bottom line," declares Kalsec, Inc., of Kalamazoo, Michigan. Its ad shows a technician standing before a computer and measuring chemicals into a test tube. The ad extols the company's "patented natural color systems." The terms "natural" and "patented" fit seamlessly together in a conceptual scheme in which there are no trade-offs and no compromises. The natural is patentable. If you think any of this is contradictory, you will not get far in the food industry.

Organic TV Dinners

As a typical American suburbanite, I can buy not just groceries but "Whole Foods" at Fresh Fields and other upscale supermarkets. I am particularly impressed by the number of convenience foods that are advertised as "organic." Of course, one might think that any food may be whole and that all foods are organic. Terms like "whole" and "organic," however, appeal to and support my belief that the products that carry these labels are less processed and more natural—closer to the family farm—than are those that are produced by multinational megacorporations, such as Pillsbury or General Foods.

My perusal of advertisements in trade magazines helped disabuse me of my belief that all-natural, organic, and whole foods are closer to nature in a substantive sense than other manufactured products. If I had any residual credulity, it was removed by an excellent cover story, "Behind the Organic-Industrial Complex," that appeared in a recent issue of the *New York Times Magazine.* The author, Michael Pollan, is shocked, shocked to find that the prepackaged microwavable all-natural organic TV dinners at his local Whole Foods outlet are not gathered from the wild by red-cheeked peasants in native garb. They are highly processed products manufactured by multinational corporations. Contrary to the impression created by advertisements, organic and other all-natural foods are often fabricated by the same companies—using comparable technologies—as those that produce Velveeta and Miracle Whip. And the ingredients come from as far away as mega-farms in Chile—not from local farmers' markets.

Reformers who led the organic food movement in the 1960s wished to provide an alternative to agribusiness and to industrial food production, but some of these reformers bent to the inevitable. As Pollan points out, they became multimillionaire executives of Pillsbury and General Mills in charge of organic food production systems. This makes sense. A lot of advanced technology is needed to produce and market an all-natural or an organic ready-to-eat meal. Consumers inspect food labels to ward off artificial ingredients; yet they also want the convenience of a low-priced, pre-prepared, all-natural dinner.

At General Mills, as one senior vice president, Danny Strickland, told Pollan, "Our corporate philosophy is to give consumers what they want with no trade-offs." Pollan interprets the meaning of this statement as follows. "At General Mills," Pollan explains, "the whole notion of objective truth has been replaced by a value-neutral consumer constructivism, in which each sovereign shopper constructs his own reality." Mass-marketed organic TV dinners do not compromise; they combine convenience with a commitment to the all-natural, eco-friendly, organic ideology. The most popular of these dinners are sold by General Mills through its subsidiary, Cascadian Farms. The advertising slogan of Cascadian Farms, "Taste You Can Believe In," as Pollan observes, makes no factual claims of any sort. It "allows the consumer to bring his or her personal beliefs into it," as the vice president for marketing, R. Brooks Gekler, told Pollan. The absence of any factual claim is essential to selling a product, since each consumer buys an object that reflects his or her particular belief system.

What is true of marketing food is true of virtually every product. A product will sell if it is all-natural and eco-friendly and, at the same time, offers the consumer the utmost in style and convenience. A recent *New York Times* article, under the title, "Fashionistas, Ecofriendly and All-Natural," points out that the sales of organic food in the United States topped $6.4 billion in 1999 with a projected annual increase of 20 percent. Manufacturers of clothes and fashion accessories, such as solar-powered watches, are cashing in on the trend. Maria Rodale, who helps direct a publishing empire covering "natural" products, founded the women's lifestyle magazine *Organic Style*. Rodale told the *Times* that women want to do the right thing for "the environment but not at the cost of living well." Advances in technology give personal items and household wares an all-natural eco-friendly look that is also the last word in fashion. Consumers "don't want to sacrifice anything," Ms. Rodale told a reporter. Why should there be a trade-off between a commitment to nature and a commitment to the good life? "Increasingly there are options that don't compromise on either front."

The food industry does not sell food any more than the fashion industry sells clothes or the automobile industry sells automobiles. They sell imagery. The slogan, "Everything the consumer wants with no tradeoffs," covers all aspects of our dreamworld. Sex without zippers, children without zits, lawns without weeds, wars without casualties, and food without technology. Reality involves trade-offs and rather substantial ones. For this reason, if you tried to sell reality, your competitor would drive you out of business by avoiding factual claims and selling fantasy—whatever consumers believe in—instead. Consumers should not be confused or disillusioned by facts. They are encouraged to assume that they buy products of Nature or Creation. In view of this fantasy, how could consumers view genetic engineering with anything but suspicion?

Nature's Own Methods

Genetic engineering, with its stupendous capacity for increasing the efficiencies of food production in all departments, including flavors and colorings, raises a problem. How can genetic recombination be presented to the consumer as completely natural—as part of nature's spontaneous course—as have other aspects of food technology? A clean label would tell consumers there is nothing unnatural or inauthentic about genetically engineered products. Industry has responded in two complementary ways to this problem.

First, the food industry has resisted calls to label bioengineered products. Gene Grabowski of the Grocery Manufacturers Association, for example, worries that labeling "would imply that there's something wrong with food, and there isn't." Michael J. Phillips, an economist with the Biotechnology Industry Organization, adds that labeling "would only confuse consumers by suggesting that the process of biotechnology might in and of itself have an impact on the safety of food. This is not the case."

Second, manufacturers point out that today's genetic technologies do not differ, except in being more precise, from industrial processes that result in the emulsifiers, stabilizers, enzymes, proteins, cultures, and other ingredients that do enjoy the benefits of a clean label. Virtually every plant consumed by human beings—canola, for example—is the product of so much breeding, hybridization, and modification that it hardly resembles its wild ancestors. This is a good thing, too, since these wild ancestors were barely edible if not downright poisonous. Manufacturers argue that genetic engineering differs from conventional breeding only because it is more accurate and therefore changes nature less.

For example, Monsanto Corporation, in a recent full-page ad, pictures a bucolic landscape reminiscent of a painting by Constable. The headline reads, "FARMING: A picture of the Future." The ad then represents genetic engineering as all natural—or at least as natural as are conventional biotechnologies that have enabled humanity to engage successfully in agriculture. "The products of biotechnology will be based on nature's own methods," the ad assures the industry. "Monsanto scientists are working with nature to develop innovative products for farmers of today, and of the future."

In this advertisement, Monsanto applies the tried-and-true formula to which the food industry has long been committed—presenting a technology as revolutionary, innovative, highly advanced, and as "based on nature's own methods." *Everything* is natural. Why not? As long as there are no distinctions, there are no trade-offs. Consumers can buy what they believe in. A thing is natural if the public believes it is. "There is something in this more than natural," as Hamlet once said, "if philosophy could find it out."

Four Concepts of the Natural

If consumers reject bioengineered food as "unnatural," what does this mean? In what way are foods that result from conventional methods of genetic mutation and selection, which have vastly altered crops and livestock, more "natural" than those that depend in some way on gene splicing? Indeed, is anything in an organic TV dinner "natural" other than, say, the rodent droppings that may be found in it? Since I am a philosopher, not a scientist, I am particularly interested in the moral, aesthetic, and cultural—as distinct from the chemical, biological, or physical—aspects of the natural world. I recognize that many of us depend in our moral, aesthetic, and spiritual lives on distinguishing those things for which humans are responsible from those that occur as part of nature's spontaneous course.

Philosophers have long pondered the question whether the concept of the natural can be used in a normative sense—that is, whether to say that a practice or a product is "natural" is somehow to imply that it is better to that extent than one that is not. Why should anyone assume that a product that is "natural" is safer, more healthful, or more aesthetically or ethically attractive than one that is not? And why is technology thought to be intrinsically risky when few of us would survive without quite a lot of it?

Among the philosophers who have questioned the "naturalistic fallacy"—the assumption that what is natural is for that reason good—the nineteenth-century British philosopher John Stuart Mill has been particularly influential. In his "Essay on Nature," Mill argues that the term "nature" can refer either to the totality of things ("the sum of all phenomena, together with the causes which produce them") or to those phenomena that take place "without the agency . . . of man." Plainly, everything in the world—including every technology—is natural and belongs equally to nature in the first sense of the term. Mill comments:

> To bid people to conform to the laws of nature when they have no power but what the laws of nature give them—when it is a physical impossibility for them to do the smallest thing otherwise than through some law of nature—is an absurdity. The thing they need to be told is, what particular law of nature they should make use of in a particular case.

Of nature in the second sense—that which takes place without the agency of man—Mill has a dour view. "Nearly all the things which men are hanged or imprisoned for doing to one another, are nature's every day performances," Mill wrote. Nature may have cared for us in the days of the Garden of Eden. In more recent years, however, humanity has had to alter Creation to survive. Mill concludes, "For while human action cannot help conforming to nature in one meaning of the term, the very aim and object of action is to alter and improve nature in the other meaning."

Following Mill, it is possible to distinguish four different conceptions of nature to understand the extent to which bioengineered food may or may not be natural. These four senses of the term include:

1) **Everything in the universe.** The significant opposite of the "natural" in this sense is the "supernatural." Everything technology produces has to be completely natural because it conforms to all of nature's laws and principles.
2) **Creation in the sense of what God has made.** The distinction here lies between what is sacred because of its pedigree (God's handiwork) and what is profane (what humans produce for pleasure or profit).
3) **That which is independent of human influence or contrivance.** The concept of "nature" or the "natural" in this sense, e.g., the "pristine," is understood as a privative notion defined in terms of the absence of the effects of human activity. The opposite of the "natural" in this sense is the "artificial."
4) **That which is authentic or true to itself.** The opposite of the "natural" in this sense is the specious, illusory, or superficial. The "natural" is trustworthy and honest, while the sophisticated, worldly, or contrived is deceptive and risky.

These four conceptions of nature are logically independent. To say that an item or a process—genetic engineering, for example—is "natural" because it obeys the laws of nature, is by no means to imply it is "natural" in any other sense. That genetically manipulated foods can be found within 1) the totality of phenomena does not show that they are "natural" in the sense that they are 2) part of primordial Creation; 3) free of human contrivance; or 4) authentic and expressive of the virtues of rustic or peasant life.

The problem of consumer acceptance of biotechnology arises in part because the food industry sells its products as natural in the last three senses. The industry wishes to be regulated, however, only in the context of the first conception of nature, which does not distinguish among phenomena on the basis of their histories, sources, or provenance. The industry argues that only the biochemical properties of its products should matter to regulation; the process (including genetic engineering) is irrelevant to food safety and should not be considered.

The food industry downplays the biochemical properties of its products, however, when it advertises them to consumers. The industry—at least if the approach taken by General Mills is typical—tries to give the consumer whatever he believes in. If the consumer believes in a process by which rugged farmers on the slopes of the Cascades raise organic TV dinners from the soil by sheer force of personality, so be it. You will see the farm pictured on the package to suggest the product is close to Creation, free of contrivance, and authentic or expressive of rural virtues. What you will not see on any label—if the industry has its way—is a reference to genetic engineering. The industry believes regulators should concern themselves only with the first concept of nature—the scientific concept—and thus with the properties of the product. Concepts related to the process are used to evoke images that "give consumers what they want with no trade-offs."

Shakespeare on Biotechnology

I confess that, as a consumer, I find organic foods appealing and I insist on "all-natural" ingredients. Am I just foolish? You might think that I would see through labels like "all natural" and "organic"—not to mention "whole" foods—and that I would reject them as marketing ploys of a cynical industry. Yet like many consumers, I want to believe that the "natural" is somewhat better than the artificial. Is this just a fallacy?

Although I am a professional philosopher (or perhaps because of this), I would not look first to the literature of philosophy to understand what may be an irrational—or at least an unscientific—commitment to buying "all natural" products. My instinct would be to look in Shakespeare to understand what may be contradictory attitudes or inexplicable sentiments.

Shakespeare provides his most extensive discussion of biotechnology in *The Winter's Tale,* one of his comedies. In Act IV, Polixenes, King of Bohemia, disguises himself to spy upon his son, Florizel, who has fallen in love with Perdita, whom all believe to be a shepherd's daughter. In fact, though raised as a shepherdess, Perdita is the castaway daughter of the King of Sicily, a close but now estranged friend of Polixenes. Perdita welcomes the disguised Polixenes and an attendant lord to a sheep-shearing feast in late autumn, offering them dried flowers "that keep / Seeming and savour all winter long." Polixenes merrily chides her: "well you fit our ages/With flowers of winter."

She replies that only man-made hybrids flourish so late in the fall:

> . . . carnations, and streak'd gillyvors,
> Which some call nature's bastards. Of that kind
> Our rustic garden's barren; and I care not
> To get slips of them.

Polixenes asks why she rejects cold-hardy flowers such as gillyvors, a dianthus. She answers that they come from human contrivance, not from "great creating nature." She complains there is "art" in their "piedness," or variegation. Polixenes replies: "Say there be;

> Yet nature is made better by no mean
> But nature makes that mean; so over that art
> Which you say adds to nature, is an art
> That nature makes. . . . This is an art
> Which does mend nature—change it rather; but
> The art itself is nature.

The statement, "The art itself is nature" anticipates the claim made by Monsanto that "The products of biotechnology will be based on nature's own methods." Polixenes, Mill, and Monsanto remind us that everything in the universe conforms to nature's own principles, and relies wholly on nature's powers. From a scientific perspective, in other words, all nature is one. The mechanism of a lever, for example, may occur in the physiology of a wild animal or in the structure of a machine. Either way, it is natural. One might be forced to agree, then, that genetic engineering applies nature's own methods and principles; in other words, "the art itself is nature."

The exchange between Perdita and Polixenes weaves together the four conceptions of nature I identified earlier in relation to John Stuart Mill. When Polixenes states, "The art itself is nature," he uses the term "nature" to comprise everything in the Universe, that is, everything that conforms to physical law. Second, Perdita refers to "great creating nature," that is, to Creation, i.e., the primordial origin and condition of life before the advent of human society. Third, she contrasts nature to art or artifice by complaining that hybrids do not arise spontaneously but show "art" in their "piedness." Finally, Perdita refers to her "rustic garden," which, albeit cultivated, is "natural" in the sense of simple or unadorned, in contrast to the ornate horticulture that would grace a royal garden. The comparison between the court and the country correlates, of course, with the division that exists in Perdita herself—royal in carriage and character by her birth, yet possessed of rural virtues by her upbringing.

Shakespeare elaborates this last conception of "nature" as the banter continues between Perdita and the disguised Polixenes. To his assertion, "The art itself is nature," Perdita concedes, "So it is." Polixenes then drives home his point: "Then make your garden rich in gillyvors,/ And do not call them bastards."

To which Perdita responds:

> I'll not put
> The dibble in earth to set one slip of them;
> No more than were I painted I would wish
> This youth should say 'twere well, and only therefore
> Desire to breed by me.

Besides comparing herself to breeding stock—amusing in the context, since she speaks to her future father-in-law in the presence of his son—Perdita reiterates a fourth and crucial sense of the "natural." In this sense, what is "natural" is true to itself; it is honest, authentic, and genuine. This conception reflects Aristotle's theory of the "nature" of things, which refers to qualities that are spontaneous because they are inherent or innate.

Perdita stands by her insistence on natural products—from flowers she raises to cosmetics she uses—in spite of Polixenes' cynical but scientific reproofs. Does this suggest Perdita is merely a good candidate for Ms. Rodale's organic chic? Should she receive a free introductory copy of *Organic Style*? Certainly not. There is something about Perdita's rejection of biotechnology that withstands this sort of criticism. Why have Perdita's actions a moral authority or authenticity that the choices consumers make today may lack?

Having It Both Ways

Perdita possesses moral authority because she is willing to live with the consequences of her convictions and of the distinctions on which they are based. By refusing to paint herself to appear more attractive, for example, Perdita contrasts her qualities, which are innate, to those of the "streak'd gillyvor," which owe themselves to technological meddling. This comparison effectively gives her the last word because she suits the action to it: she does not and would not paint herself to attract a lover. Similarly, Perdita does not raise hybrids, though she admits, "I would I had some flow'rs" that might become the "time of day" of the youthful guests at the feast, such as Florizel.

Perdita does not try to have it both ways—to reject hybrids but also to grow cold-hardy flowers. She ridicules those who match lofty ideals with ordinary actions—whose practice belies their professed principles. For example, Camillo, the Sicilian lord who attends Polixenes, compliments Perdita on her beauty. He says, "I should leave grazing, were I of your flock,/And only live by gazing." She laughs at him and smartly replies, "You'd be so lean that blasts of January/Would blow you through and through."

Many people today share Perdita's affection for nature and her distaste for technology. Indeed, it is commonplace to celebrate Nature's spontaneous course and to condemn the fabrications of biotechnology. Jeremy Rifkin speaks of "Playing Ecological Roulette with Mother Nature's Designs"; Ralph Nader has written the foreword to a book titled, *Genetically Engineered Food: Changing the Nature of Nature.* The Prince of Wales, in a tirade against biotechnology, said, "I have always believed that agriculture should proceed in harmony with nature, recognising that there are natural limits to our ambitions. We need to rediscover a reverence for the natural world to become more aware of the relationship between God, man, and creation."

While consumers today share Perdita's preference for the natural in the sense of the authentic and unadorned and spurn technological meddling, they do not share her willingness to live with the consequences of their commitment. They expect to enjoy year-round fruits and vegetables of unblemished appearance, and consistent taste and nutritional quality. Gardeners wish to plant lawns and yards with species that are native and indigenous, and they support commissions and fund campaigns to throw back the "invasions" of exotic and alien species. Yet they also want lawns that resist drought, blight, and weeds, and—to quote Perdita again—to enjoy flowers that "come before the swallow dares, and take/The winds of March with beauty." In other words, the consumer wants it both ways. Today's consumers, as Ms. Rodale knows, "don't want to sacrifice anything." Today's consumers insist, as did Perdita, on the local, the native, the spontaneous. Yet they lack her moral authority because they are unwilling to live with the consequences of their principles or preferences. Consumers today refuse to compromise; they expect fruits and flowers that survive "the birth/Of trembling winter" and are plentiful and perfect all year round.

Naked Lunch

Those who defend genetic engineering in agriculture are likely to regard as irrational consumer concerns about the safety of genetically manipulated crops. The oil and other products of Roundup Ready soybeans, according to this position, pose no more risks to the consumer than do products from conventional soybeans. Indeed, soybean oil, qua oil, contains neither DNA nor protein and so will be the same whether or not the roots of the plant are herbicide resistant. Even when protein or DNA differs, no clear argument can be given to suppose that this difference—e.g., the order of a few nucleotides—involves any danger. Crops are the outcome of centuries or millennia of genetic crossing, selection, mutation, breeding, and so on. Genetic engineering adds but a wrinkle to the vast mountains of technology that separate the foods we eat from wild plants and animals.

The same kind of argument may undermine consumer beliefs that "natural" colors and flavors are safer or more edible than artificial ones. In fact, chemical compounds that provide "natural" and "artificial" flavors can be identical and may be manufactured at the same factories. The difference may lie only in the processes by which they are produced or derived. An almond flavor that is produced artificially, as I have mentioned, may be purer and therefore safer than one extracted from peach or apricot pits. Distinctions between the natural and the artificial, then, need not correspond with differences in safety, quality, or taste—at least from the perspective of science.

Distinctions consumers draw between the natural and the artificial—and preferences for the organic over the engineered—reflect differences that remain important nonetheless to our cultural, social, and aesthetic lives. We owe nature a respect that we do not owe technology. The rise of objective, neutral, physical and chemical science invites us, however, to disregard all such moral, aesthetic, and cultural distinctions and act only on facts that can be scientifically analyzed and proven. Indeed, the food industry, when it is speaking to regulators rather than advertising to consumers, insists on this rational, objective approach.

In an essay titled, "Environments at Risk," Mary Douglas characterizes the allure of objective, rational, value-neutral, science:

> This is the invitation to full self-consciousness that is offered in our time. We must accept it. But we should do so knowing that the price is William Burroughs' *Naked Lunch.* The day when everyone can see exactly what it is on the end of everyone's fork, on that day there is no

> pollution and no purity and nothing edible or inedible, credible or incredible, because the classifications of social life are gone. There is no more meaning.

Advances in genetic engineering invite us to the full self-consciousness that Douglas describes and aptly analogizes to the prison life depicted in *Naked Lunch.* It is the classifications of social life—not those of biological science—that clothe food and everything else with meaning. Genetic engineering poses a problem principally because it crosses moral, aesthetic, or cultural—not biological—boundaries. The fact that the technology exists and is successful shows, indeed, that the relevant biological boundaries (i.e., between species) that might have held in the past now no longer exist.

Given advances in science and technology, how can we maintain the classifications of social life—for example, distinctions between natural and artificial flavors and between organic and engineered ingredients? How may we, like Perdita, respect the difference between the products of "great creating nature" and those of human contrivance? Perdita honors this distinction by living with its consequences. Her severest test comes when Polixenes removes his disguise and threatens to condemn her to death if she ever sees Florizel again. Florizel asks her to elope, but she resigns herself to the accident of their origins—his high, hers (she believes) low—that separates them forever. Dressed up as a queen for the festivities, Perdita tells Florizel: "I will queen it no further. Leave me, sir; I will go milk my ewes and weep."

Perdita, of course, both renounced her cake and ate it, too. In Act IV, she gives up Florizel and his kingdom, but in Act V she gets them. Her true identity as a princess is eventually discovered, and so the marriage happily takes place. If you or I tried to live as fully by our beliefs and convictions—if we insisted on eating only those foods that come from great creating nature rather than from industry—we would not be so fortunate. "You'd be so lean that blasts of January/Would blow you through and through."

Perdita is protected by a playwright who places her in a comedy. Shakespeare allows her to live up to her convictions without compromising her lifestyle. This is exactly what the food industry promises to do—"to give consumers what they want with no trade-offs." It is exactly what Ms. Rodale offers—to protect the environment "but not at the cost of living well." The food, fashion, and other industries work off stage to arrange matters so that consumers can renounce genetic engineering, artificial flavors, industrial agriculture, and multinational corporations. At the same time, consumers can enjoy an inexpensive, all-natural, organic, TV dinner from Creation via Cascadian Farms.

Perdita lives in the moral order of a comedy. In that moral order, no compromises and no trade-offs are necessary. You and I are not so fortunately situated. Indeed, we must acknowledge the tragic aspect of life —the truth that good things are often not compatible and that we have to trade off one for the sake of obtaining the other. The food industry, by suggesting that we can have everything we believe in, keeps us from recognizing that tragic truth. The industry makes all the compromises and hides them from the consumer.

This article is based on a presentation made at the National Agricultural Biotechnology Council's annual meeting, "High Anxiety and Biotechnology: Who's Buying, Who's Not, and Why?," held May 22–24, 2001. A version of this article is forthcoming in NABC Report 13 symposium proceedings.

The author acknowledges the support of the National Human Genome Research Institute program on Ethical, Legal, and Social Implications of Human Genetics, Grant R01HG02363; also the National Science Foundation, Grant 9729295.

Sources

Food Processing, February 1988. Lucy Saunders, "Selecting an Enzyme," *Food Product Design* (May 1995), online at: http://www.foodproductdesign.com/archive/1995/0595AP.html; Michael Pollan, "Behind the Organic-Industrial Complex," *New York Times Magazine* (May 13, 2001); Ruth La Ferla, "Fashionistas, Ecofriendly and All-Natural," *New York Times* (July 15, 2001); Bill Lambrecht, "Up to 50%+ of Crops Now Genetically Modified," *St. Louis Post-Dispatch,* Washington Bureau (August 22, 1999), and available online at http://www.healthresearchbooks.com/articles/labels2.htm; Jim Wilson, "Scientific Food Fight," in *Popular Mechanics* online, which is available at http://popularmechanics.com/popmech/sci/0002STRSM.html; John Stuart Mill, "Nature" in *Three Essays on Religion* (New York, Greenwood Press, 1969), reprint of the 1874 ed.; Jeremy Rifkin, "The Biotech Century: Playing Ecological Roulette with Mother Nature's Designs," in *E Magazine* (May/June 1998); Martin Teitel and Kimberly A. Wilson, *Genetically Engineered Food: Changing the Nature of Nature: What You Need to Know to Protect Yourself, Your Family, and Our Planet* (Vermont: Inner Traditions, Int'l, Ltd., 1999); "Seeds of Disaster: An Article by The Prince of Wales," *Daily Telegraph* (June 8, 1998); Mary Douglas, "Environments at Risk," in *Implicit Meanings: Essays in Anthropology* (London: Routledge & Kegan Paul, 1975).

Unnatural Farming and the Debate over Genetic Manipulation

Paul B. Thompson

The suggestion that genetics involves forbidden or toxic knowledge points toward a certain queasiness that many feel in connection with the new science of genomics—the scientific discipline of mapping, sequencing, and analyzing an organism's genetic material. The frequent use of Frankenstein metaphors, as well as talk about "playing God," also indicate trouble. The source of the trouble is sometimes expressed by saying that there is something "unnatural" about the research and technical applications involving recombinant DNA. The most obvious ways to interpret this queasiness are to see the unnatural either as a religious taboo or as a cultural construction. In either case, showing the operative understandings of nature and the natural to be ill-considered, inconsistent, and poorly informed is a fairly straightforward project that has been successfully completed a number of times.

When products are banned or discouraged on the ground that they are unnatural, the record suggests that this rationale is relied on because no good reason can be produced. One is left with two choices. One can abandon altogether arguments that make a normative appeal to nature. Alternatively, one can accept some form of the argument from repugnance made famous by the medical ethicist Leon Kass in his work on cloning; that is, the widespread reaction of queasiness itself gives a basis for the suspicion that there is more here than meets the eye. Even if we cannot say exactly why we find this domain of science to be disturbing, we are well advised to pay a certain heed to these feelings.

In the years since Ian Wilmut, Keith Campbell, and their colleagues at the Roslyn Institute announced the successful cloning of the sheep, Dolly, much of the public outcry over biotechnology has centered on genetically engineered agricultural crops. Empirical research on the European public indicates that, among those who oppose genetic engineering in agriculture and food, opposition is based on the belief that the technology is unnatural. Respondents to surveys do not tend to believe that genetically engineered foods are unsafe to eat, but that the food crops are harmful to the environment. What is more, there is a strong correlation between objections to genetically modified organisms (GMOs) and the possibility of human cloning, though there is less concern with other applications of biotechnology in human medicine. Nevertheless, the uproar over agricultural genetic engineering has led some to speculate that public outrage over GMOs may spill over and form more widespread opposition to biotechnology.

Agricultural Biotechnology and the Natural

Mark Sagoff, in "Genetic Engineering and the Concept of the Natural," uses the controversy over GMOs to analyze the belief that products of biotechnology are unnatural. Sagoff chides the food industry for promoting in its advertising an incoherent and unachievable ideal of natural goodness, while at the same time opposing reference to nature or naturalness in regulation of GMOs. Sagoff's larger philosophical point is that the aesthetic and cultural conceptions of nature and the natural that inform our habits, mores, and consumer preferences have no basis in any biological difference between GMOs and other agricultural products. In making this argument, he relies on John Stuart Mill's analysis of four senses in which something can be called "natural." In one sense, nature is simply the totality of existing things, everything in the universe. In another, it is God's handiwork, as distinct from man's, while in a third, it is that which is independent of human influence or contrivance. The fourth way in which something can be natural is in the sense of being authentic or true to itself. Sagoff's argument in a nutshell is that the first three senses do not provide the basis for a biological or normative distinction between GMOs and conventional agricultural crops, which are the product not only of advanced breeding programs, but also of centuries of human artifice. The final sense relies on legitimate aesthetic and cultural norms, but cannot be supported by science.

Sagoff is circumspect in the end about whether science or culture should govern policy with respect to genetically engineered crops, but he is clear in noting that we cannot have it both ways. Sagoff does not say that society ought not to ban GMOs—only that society should not look to science for reasons so to do. Furthermore, if culture is to reign, we must be prepared to give up many of the conveniences that science has brought to our food system. Although he does not specifically refer to biomedical applications of biotechnology in his analysis, Sagoff's position easily can be extended in that direction. Inconsistencies in the way that we interpret the world based on our conception of the natural could just as easily promote turmoil and outrage over medical biotechnology. Another implication of Sagoff's view is that if we are to insist *consistently* on aesthetic and cultural criteria for our normative conception of the natural, then we have arguments against the use of biotechnology in agriculture *and* medicine.

Perhaps we should simply rejoice in the empirical evidence that suggests that the public is inconsistent in its attitudes, but before doing so, I would like to complicate Sagoff's analysis of the debate over agricultural biotechnology by offering a fifth sense of the natural to Mill's four. While Sagoff may have quite accurately captured the public's confusion over natural foods, there is more to the debate over the naturalness of agricultural biotechnology than is suggested by his analysis. I offer what I will call *the artisanal conception of the natural* as it relates to natural farming. I also examine how and whether this conception of the natural has any purchase within the medical context, and I conclude with some reflections on the mutual implications of the debate over biotechnology and the natural for medical and agricultural bioethics.

Natural Farming

To some degree, the use of the word "natural" in connection with agriculture means exactly—which is to say, vaguely—just what Sagoff says it means. It evokes a romantic vision of pure essences, contaminated, polluted, and ultimately dissipated by modern technology and the rationalism of the scientific worldview. However, this romantic, inspirational and ultimately futile conception of the natural bleeds seamlessly into a more pedestrian, practical, and commonsense idea in agrarian contexts. This agrarian conception of the natural is best understood by considering what it means for a farmer to do something that is "against nature," and to sketch an analogy between farmers and other artisans who work with natural materials, such as woodcarvers or stonemasons. Artisans work against nature when they go against the grain. Wood and stone materials have a natural grain that determines physical properties such as tensile or compression strength, properties critical to cutting and shaping wood and stone. Neglecting these properties not only makes the work of the artisan more difficult, but it also can result in a weak and brittle finished object that is less suited to its purpose than the object made by a master craftsman.

Ordinary farming is similar in countless ways. Farmers must know where the sun shines and where the wind blows when planting crops or pasturing livestock. Laying out a field without attention to the slope and the properties of soil encourage erosion and result in the frequent need to re-seed and re-till. The mix of crops, farming practices, and even natural variation can produce synergies—such as the rotation of corn and beans, which increases available nitrogen in the soil, while limiting infestations of soil pests. At times natural variations can be exploited; for instance, the mix of predator and prey among insects in a farm field, and the genetic diversity of the crop itself, are two characteristics that can increase a crop's resistance to catastrophic failures.

Although full knowledge of natural variation in farming is both site and crop specific, one can generalize two features of the artisan's knowledge. Farmers as artisans are knowledgeable about synergies and, because they employ natural farming methods that promote resilience and resistance to catastrophic failure, adjustments in response to changing conditions require less deliberate action by farmers. Deviations from these methods create weak or brittle spots in the farming system, calling for farmers to devote more of their future time to perennial repair and maintenance.

Farmers, however, often choose the less natural of two production options. Pesticide use is farming "against the grain," because most pesticides kill both beneficial and pest insects. The farmer loses the services of pollinators, and when insect predators decline the farmer is hooked on further pesticide use. The addiction worsens as insect pests become resistant to chemical pesticides, resulting in the pesticide treadmill: the need to constantly use new and often more potent pesticides. These weaknesses in technology-intensive farming are well known: it was in 1948 that Aldo Leopold wrote that "agricultural science is largely a race between the emergence of new pests and emergence of new techniques for their control." But most farmers in the US rely on pesticides; they accept the brittleness and loss of synergy because they believe that the economic risks of pesticide use are more manageable than the economic risks of any alternative. Thus, for many working farmers, more natural is not always better.

Although every farm family continually debates the wisdom of abandoning relatively more natural or artisanal farming methods, the most acrimonious arguments surround agricultural policies and technologies that shape farm decision making. Some have characterized agricultural development in terms of adopting large-scale, industrial production methods—the so-called Green Revolution—while others have cautioned against the displacement of local, indigenous knowledge. When applied to developing countries, this debate does not concern farmer decision making at all, but rather considers whether technologies and policy approaches developed in the US are appropriate to the artisanal practices of peasant farmers.

In some quarters, the debate takes the form of a broad-based attack on modern agricultural science and associated technologies. Modern science is characterized as exogenous and universalizing, and hence as insensitive to the evolutionary and indigenous knowledge that accrues through trial and error to those artisans attentive to specific local characteristics. While this critique may appear like postmodernism, it is actually pre-modernism, and it is entirely rational from the perspective of the actual working farmer. In other quarters, the debate is taken up by scientists who are also quite sensitive to the synergies of complex farming systems and to the brittleness associated with many putatively successful agricultural technologies. The competing perspectives in this debate are also quite complex, though they are less likely to be characterized in terms of "natural" and "unnatural" agriculture.

Rather than attempt to summarize these perspectives, I shall note two emblematic opinions on the recent discovery that transgenic maize has been found growing in the region of Mexico considered the epicenter of genetic diversity for open pollinated maize. Ignacio Chapella, a microbial ecologist at the University of California at Berkeley, states that the presence of transgenes here threatens the genetic diversity on which future generations everywhere will depend. Michael J. Phillips, an agricultural economist who is a vice president of BIO (Biotechnology Industry Organization), the primary trade organization for biotechnology companies, sees it as proof that you can't keep a good technology down.

In this dispute, the terminology of natural agriculture is embedded in a long-running and important controversy over the importance of indigenous knowledge and artisanal practice in agricultural systems. At the same time, debate grows over the sustainability of the food system as a whole and the impact of agriculture on natural ecosystems. Concern for ecological integrity has stimulated interest in preserving indigenous knowledge and the genetic diversity associated with artisanal farming systems. Although scientifically trained participants are unlikely to emphasize the naturalness of any given farming practice, it is not surprising that these ecological concerns should become entangled with the rhetoric of naturalness.

Biotechnology and Natural Farming

Are applications of recombinant DNA (rDNA) and genomics compatible with the natural ideal of farming? This is a vexed question, and rather than answering it, I want to show *why* it is vexed. Certainly the history of political conflict over agricultural research during the 1980s and 1990s shows that biotechnology and natural farming are like oil and water, with the advocates of each at loggerheads. In 2000, the most successful applications of rDNA technology were herbicide-tolerant and insect-resistant crops intended for large-scale monocultures, and US standards for organic farming did not permit the use of genetically engineered crops.

Biotech agriculture got its start in the early years of the 1980s, when private sector scientists at several companies began development of the first generation of products using recombinant techniques. These included the Flav'r Sav'r tomato, modified for longer shelf life, the "Ice Minus" bacterium for protecting crops against frost, recombinant bovine somatotropin (Posilac) for increasing dairy yields, Round-Up Ready soybeans and a range of crops engineered to produce the *bacillus thuringiensis* (Bt) toxin as a protection against damage from some kinds of pests. Each venture was prompted by its expected profitability, and as these products began development, the companies undertaking them accumulated process patents that quickly became valuable. This, in turn, began a scramble to acquire expertise and market position in biotechnology among agricultural input firms that previously were primarily involved in the production of chemicals and pharmaceuticals. The upshot was that biotechnology became indelibly associated with for-profit research conducted by firms that had not historically been regarded as friends of either the environment or small-scale farmers.

The timing for this association could not have been worse. As historian David Danbom shows, although the history of resistance to science-based and technologically-driven change in farming systems is long, it reached a climax in the mid to late 1980s—when activists chronicled the role of agricultural research in displacing family-style farming, and agricultural scientists accepted the validity of an indictment levied twenty years earlier in Rachel Carson's *Silent Spring.* The phrase "sustainable agriculture" was fresh in the air, and many were hopeful that agricultural universities and institutes would launch environmentally-oriented programs in research grounded jointly on ecology and a form of social ecology called "farming systems" research. These hopes were in large measure disappointed, as research administrators felt compelled to staff their institutions with specialists in the new (and at that point expensive) recombinant techniques. As the 1990 report *Biotechnology's Bitter Harvest* makes abundantly clear, biotechnology came to be seen as inimical to sustainable agriculture, not so much because of its intrinsic biological capabilities as because of its impact on research priorities at institutions dedicated to serve the public interest.

By the end of the twentieth century, social organizations committed to organic foods or sustainable agriculture had solidified their resistance to genetic engineering in agriculture, concluding that developers of biotechnology lack the wisdom to intervene safely in complex natural systems. The farmer's incentives to adopt genetically modified (GM) crops are based both on direct savings derived from more cost-effective use of chemical applications as well as on a reduction of management effort needed for mid-season chemical application. As such, neither activist groups nor farmers using GM crops support a picture of biotechnology that can be characterized as pursuing an artisanal sense of natural farming.

A comprehensive meta-analysis conducted for the Council on Agricultural Science and Technology supports the conclusion that herbicide tolerant and Bt crops yield net environmental benefits. Since their study compares the use of GM crops with other approaches in industrial agriculture that would not be considered natural in the artisanal sense, it is unclear how one should regard this evidence in making an overall assessment of agricultural biotechnology.

Yet one can view the detailed mapping of plant genomes as enhancing our collective capacity to farm "with the grain," since genomics allows a better understanding of the fine grain of genetic capabilities in agricultural plants. As such, the methods associated with biotechnology might eventually permit farming approaches that are far more sensitive to the array of situational characteristics and variables important in artisanal farming. In short, just because first-generation biotechnology products seem incompatible with artisanal farming, it is far too early to conclude that no conceivable applications of biotechnology is incompatible with natural, artisanal farming.

On Agricultural and Biomedical Science

In assessing the relevance of these considerations to questions in medical bioethics, one can first note the similarities and differences in the agricultural and biomedical sciences. Both are characterized broadly as applied biology, though virtually all of the natural sciences are represented to some degree in each. Both agricultural and biomedical science are strongly normative applied sciences oriented toward advancing such ideals of human betterment as the promotion of health.

A number of scientific discoveries and techniques pioneered in agricultural or veterinary contexts have found application in medicine—especially for a number of reproductive technologies, including artificial insemination, in vitro fertilization, embryo transfer, and now cloning. Agricultural applications of plant and animal breeding allowed the emergence of genetics as a science and also for the eugenics movements of the early twentieth century. The history of agricultural precedents for controversial human medical practices suggests that agriculture is a harbinger of what is to come in medicine and in the bioethics of medical biotechnology.

Despite similarities and their common basis in the natural sciences, agricultural and biomedical research have developed strongly distinct cultures that also tend to be mutually exclusive domains of scientific activity. Most of the leading private universities in the US have medical schools and extensive research programs in medicine, but no private institutions offer graduate degrees in the full range of agricultural sciences. Relatively few scientists or research groups have made important discoveries in both areas. Philosophical and methodological differences account for part of this separation. Biomedical researchers tend to presume that all human beings are biologically alike, and therefore their findings would be universally applicable to all people. In contrast, agricultural researchers tend to conduct highly repetitive experiments intended to show that, for instance, the performance of a plant observed in Iowa would be the same if grown in Illinois.

The separation between agricultural and biomedical technology is not only a scientific truth but also a social artifact. Until recently, the US Department of Agriculture (USDA) did not even permit researchers at non-agricultural institutions to *apply* for funding, even if the research was beneficial to agriculture. Correlatively, the National Institutes of Health (NIH) typically has applied a strong human medical benefit test to research that it has funded. This difference is also reflected in the way that research discoveries are made available to clients. Agricultural science developed a strong culture of public funding and dissemination of research results through state extension services. Cooperation between researchers at public and private institutions has periodically sparked waves of controversy. For example, the American Society of Agronomy publicly censured Donald Jones, a researcher at the Connecticut Agricultural Experiment Station, in 1956 because he applied for a patent on his invention of a method for achieving cytoplasmic male sterility for hybrid corn. The culture of agricultural research at that time militated against deriving financial return from one's investment. Biomedical researchers, in contrast, have historically been able to work closely with drug, instrument, or medical supply companies without fear of controversy.

Although agricultural *technology* might predict future medical applications, agricultural *controversy* is not a good predictor of disagreements in medicine. Reproductive techniques such as implantation of embryos and the use of surrogate mothers were never particularly controversial when they were pioneered on agricultural animals. Yet ethical debate arose when these technologies were first used to counter human infertility. All this suggests that despite their scientific similarities, the conceptual gap between agriculture and medicine continues to loom large in the public's mind. History provides little basis for thinking that just because biotechnology has been thought unnatural in agriculture, this will carry over to controversy in medicine.

Ethical Implications for Medicine and Bioethics

Even if there is little reason to think that an agricultural genetics controversy will precipitate a medical one, several ethical issues remain worth considering. One can ask at least three questions about the relation of the agrarian or artisanal conception of natural agriculture to medical biotechnology and standard medical ethics. First, we must consider whether the legitimate questions associated with the agricultural debate have any analog in medicine and medical ethics. We must also consider how spillover from the agricultural debate might influence the way that people interpret the naturalness of human biotechnology in a medical context. Finally, we must consider how the biomedical debate over the naturalness of biotechnology, cloning, and new genetically based medical technology might affect agricultural issues.

Analogies can be drawn between the indigenous medical knowledge of shamans and healers and indigenous farming practices. Legitimate concerns arise about the possibility that modern scientific medicine will erase this knowledge base. Yet it seems equally clear on its face that the place of natural artisanal medicine is quite different from that of natural agriculture in the pantheon of bioethics issues. The transformation and industrialization of farming is arguably *the* central question in agricultural ethics. One's view on this question also places one in a number of intellectually and politically entrenched camps. The most that I would be prepared to say is that the debate about indigenous farming in agriculture should at least prompt thought among medical bioethicists about indigenous medical knowledge. Interesting questions can also be raised about the relationship between medical science and medical practice. As the philosopher of science Frederick Suppe has argued, it is clear that place-based, particularized knowledge is fundamental to farming, and that successful farming will never be a direct application of the scientific findings of agricultural science. Even the most high-tech farming is thus an adaptation of science to a farming mentality that still has its roots in the trial and error wisdom of farming with the natural grain. Perhaps something similar could be said about medical practice.

The substantive intellectual distance between natural agriculture's importance for the debate on agricultural biotechnology, on the one hand, and natural medicine's importance for genetic technologies applied to human beings, on the other, suggests most of what is worth noting about spillover. First, just because some find agricultural biotechnology highly unnatural, we should not expect them to raise the same concern about medical biotechnology. Medical biotechnologies are not unnatural in the sense of "going against the grain" and forcing doctors to abandon wisdom acquired over generations of practice. We do not think of medical biotechnologies as being foisted on us without a choice, and even if we did, we would not associate this problem with their naturalness. All these considerations suggest that these issues will remain separate.

However, the intellectual distance between the ethical problems of agricultural and medical biotechnology also suggests that if there *is* spillover, it would be the worst kind of spillover one could imagine. Rather than being an instance of analogous problems, it would be a case of moral rhetoric run amok and being applied almost wholly out of context. Concretely, the scenario might look like this. People see respected professors of ecology and representatives of nongovernmental organizations (NGOs) expressing concern about the ecological impact of agricultural biotechnology. They interpret this concern as meaning that biotechnology is against nature, is unnatural. Leaping chasms of logic, they apply their concern to medical genetics or new therapies, now perhaps thinking of "unnatural" more in the sense of impure or polluting. The public rises up in protest, creating political and legal hurdles for important medical technology. Although this is not a likely scenario, it may be the worst case, suggesting the value in taking pains to block it. As when, for instance, Friends of the Earth, historically dedicated to environmental issues, declared its opposition to the prospect of human cloning by insisting that cloning is contrary to "the principle of respect for nature."

But blocking this kind of spillover could have disastrous impact on the agricultural debate, and ultimately on agriculture itself. Here is how that scenario goes: Fearing public backlash, respected biomedical scientists and leading bioethicists make blanket statements condemning criticisms of genetic technology and pooh-poohing anyone who suggests that such technologies are unnatural. This discredits the ecology professors and plays into the hands of industry pimps who are trying to use gene technology in a larger ploy to chase poor peasant farmers from their lands in Africa, Asia, and Latin America, so that industrial growers—who will purchase not only genetically engineered seeds, but also tractors, harvesters, and agrochemicals—can take their place. Food prices stay low, but poverty spreads throughout the developing world, nevertheless. What is more, important centers of genetic diversity are lost over the next generation. In 2050, a new mutant virus attacks a gene locus that happens to have been bred into many crops and through ingression is now almost ubiquitous in the world's five major crops. Everyone starves.

This is obviously a slippery slope argument that characterizes the morality of actors in caricaturist fashion. Nevertheless, this kind of reverse spillover is the most serious risk we face as medical ethics debunks the idea of the natural. Important and substantive issues are debated in agricultural biotechnology and the rhetoric of naturalness cannot easily be extricated from that debate. It would thus be unfortunate for philosophers to debunk all uses of the concept of the natural. It *would* be helpful for medical ethicists to competently speak to the most serious issues of agriculture—especially when, in issues involving livestock or plants, journalists are likely to rely on the opinions of medical ethicists than to seek out those knowledgeable about agriculture.

Conclusion

Although much of the public's worry about the unnatural character of plant and animal biotechnology is undoubtedly based on precisely the muddles that Sagoff diagnoses in terms of Mill's four-part schema, substantive debates over relatively more and less natural farming methods suggest a fifth sense of natural. This sense is based on the detailed and place-specific knowledge that farmers derive from following a farming strategy for many years, sometimes for generations. It is a sense of the natural that harks back to the practice of artisans who find synergy and avoid brittleness by attention to the natural properties of their materials. A debate over the loss of indigenous knowledge has raged in agriculture for at least a century. While some antagonists have pitted science against the artisanal conception of the natural, others have sought to incorporate this knowledge into agricultural science itself. This debate has entered a new phase with respect to biotechnology.

Broadly speaking, we should endorse a view of agriculture that is attentive to the details that make farming in one place quite different from farming someplace else. Forces that tend to impose a single set of standards on farming practices or that presume universal applicability of technology weaken the sustainability of our food system. Furthermore, as argued above, genomics and genetic engineering figure in the debate in a complex, uncertain, and certainly contested manner. Some will see this new science as inimical to the detailed kind of artisanal practice associated with natural farming, while others will see the genetic fine structure revealed by this science as offering important new avenues for farming "with nature." In either case, those who hold that one cannot meaningfully apply a conception of the natural in a normative argument need to be cognizant of this debate about our agricultural practice. Whether there are more substantive issues in medical bioethics where the artisanal conception of the natural might profitably be applied remains an open question.

Sources

Leon R. Kass, "The Wisdom of Repugnance," *The New Republic* (June 2, 1997); I. A. E. Wilmut, A. E. Schnicke, J. McWhir, A. J. Kind, and K. H. S. Campbell, "Viable Offspring Derived from Fetal and Adult Mammalian Cells," *Nature,* vol. 385 (1997); on empirical research showing that the European public views technology as unnatural, see John Durant, Martin W. Bauer and George Gaskell, *Biotechnology in the Public Sphere: A European Sourcebook* (The Science Museum, 1998) and also Wolfgang Wagner, Nicole Kronberger, et al., "Pandora's Genes—Images of Genes and Nature," in *Biotechnology: The Making of a Global Controversy,* edited by M. W. Bauer and G. Gaskell (Cambridge University Press, 2002); for speculation that outrage over GMOs may lead to widespread opposition to biotechnology generally, see Brian Tokar, "Challenging Biotechnology," in *Redesigning Life? The World Challenge to Genetic Engineering,* edited by Brian Tokar (Zed Books, 2001), Richard Lewontin, "Genes in the Food!" *New York Review of Books,* vol. 48 (2001), Brian Salter and Mavis Jones, "Human Genetic Technologies, European Governance and the Politics of Bioethics," *Nature Reviews/Genetics,* vol. 3 (2002); Jochen Bockmühl, "A Goethean View of Plants: Unconventional Approaches," in *Intrinsic Value and Integrity of Plants in the Context of Genetic Engineering,* edited by D. Heaf and J. Wirz (International Forum for Genetic Engineering, 2001); Aldo Leopold, *A Sand County Almanac and Sketches Here and There* (Oxford University Press, 1949); for a caution against displacement of local indigenous knowledge, see John W. Bennett, "Research on Farmer Behavior and Social Organization," in *New Directions for Agriculture and Agricultural Research,* edited by K. A. Dahlberg (Rowman and Allenheld, 1986); for a discussion of US policy applied to artisanal practices of peasant farmers, see David Barkin, "Sustainability: The Political Economy of Autonomous Development," *Organization and Environment,* vol. 11 (1998); Ignacio Chappella and Michael Phillips were interviewed for a segment on *Bill Moyers NOW* (originally broadcast October 4, 2002); on the history of biotech agriculture, see Martin Kenney, *Biotechnology: The University-Industrial Complex* (Yale University Press, 1986) and also David B. Danbom, *The Resisted Revolution: Urban America and the Industrialization of Agriculture, 1900-1930* (Iowa State University Press, 1979); on the role of agricultural research in displacing family style farming, see Nancy Shankle, "Agricultural Research Policy and the Family Farm," in *Beyond the Large Farm: Ethics and Research Goals for Agriculture,* edited by P. B. Thompson and B. A. Stout (Westview Press, 1991); Rachel Carson, *Silent Spring* (Houghton Mifflin, 1962); L.

Busch, W. Lacy, J. Burkhardt, L. Lacy, *Plants, Power and Profit: Social, Economic and Ethical Consequences of the New Biotechnologies* (Basil Blackwell, 1991); Rebecca Goldberg, Jane Rissler, Hope Shand, and Chuck Hassebrook, *Biotechnology's Bitter Harvest* (Biotechnology Working Group, 1991); Ann Elizabeth Reisner, "Social Movement Organizations' Reactions to Genetic Engineering in Agriculture," *American Behavioral Scientist*, vol. 44 (2001); David S. Bullock and Elisavet I. Nitsi, "Round-Up Ready Soybean Technology and Farm Production Costs: Measuring the Incentive to Adopt Genetically Modified Seeds," *American Behavioral Scientist*, vol. 44 (2001); Janet Carpenter, Allan Felsot, Timothy Goode, Michael Hammig, David Onstad, and Sujatha Sankula, *Comparative Environmental Impacts of Biotechnology-Derived and Traditional Soybean, Corn, and Cotton Crops* (Council for Agricultural Science and Technology, Ames, Iowa, 2002); on the prospect that biotechnology might become more sensitive and permit artisanal farming, see Robert Goodman, "Ensuring the Scientific Foundations for America's Agriculture," in *Visions of American Agriculture*, edited by W. Lockeretz (Iowa State University Press, 1997) and Paul B. Thompson, "The Environmental Ethics Case for Crop Biotechnology: Putting Science Back into Environmental Practice," in *Moral and Political Reasoning in Environmental Practice*, edited by A. Light and A. De-Shalit (MIT Press, 2003); Frederick Suppe, "The Limited Applicability of Agricultural Research," *Agriculture and Human Values*, vol. 4 (1987); David MacKenzie, "Agroethics and Agricultural Research," in *Beyond the Large Farm: Ethics and Research Goals for Agriculture*, edited by P. B. Thompson and B. A. Stout (Westview Press, 1991); Testimony of Dr. Brent Blackwelder, President of Friends of the Earth before the Senate Appropriations Committee, Concerning the Cloning of Humans and Genetic Modifications, January 24, 2002, and available at: http://www.foe.org/site1/act/testimonycloning.html.

Our (Modified?) Human Nature

II

Genetic Engineering and Our Human Nature

Harold W. Baillie

Introduction

Developments in genetic engineering have swept us along faster than we can follow—and certainly faster than allows for adequate consideration of ethical consequences. Genetic engineering in plants and animals is pervasive, human genetic selection is now commonplace, and genetic enhancement seems unavoidable. Challenging and dangerous possibilities of "improving" the species will, discovery by discovery, arrive at our threshold like the salesman, who, his foot initially thrust in the door, is suddenly standing in the foyer. Given the pace of change and our fascination with it, one can reasonably anticipate that arguments against genetic engineering will be ignored in the face of technological enthusiasm. For better or worse, our children will receive from us a brave new world.

If we are "condemned" to bequeath our children a world in which genetic engineering is a reality, we must develop a way of talking about—and enforcing—rules governing its use in a way that furthers human concerns while preserving human nature. In short, we must assure that, in some sense, our children are like us. But up to now discussions about genetic engineering have been inadequate. There is a vague insistence on the need for "enlightened" public policy, both national and international, to ensure a harmonious world community.

But policies addressing genetic engineering must reflect some vision of what human beings *are,* a task that still lacks a consensus. If we are to preserve our human nature, perhaps we ought first consider the question: what *have we been* as human beings? I want to suggest that the notion of the sacred is an important part of such considerations. Understanding the sacred helps us both to identify elements in nature and human nature that ought to be preserved and also to understand what might be required to "preserve" our human nature.

A Traditional Notion of the Sacred

The notion of the sacred refers to that which is holy or has been made holy by its connection with a deity. For example, the medieval theologian Thomas Aquinas uses the term to describe the science of revelation; for him, the sacred refers to both God and His actions, and subsequently to what has been revealed to us about them. Revelations of the sacred is for our benefit, and although human understanding is limited—a weakness of our intellect, contends Aquinas—revelation is an incitement and opportunity to improve our wisdom about the nature of the world and our place in it.

According to Aquinas, humans are created in the "image and likeness of God." The significance of this is that an image carries the sign of its origin and an aspect of the original. In humans, the image of God is found in the "intellectual soul," which houses reason and the possibility of judgment. The rest of the animal world carries a "trace" rather than an image of God—that is, evidence of God and His design, but not any element that allows for reason and judgment.

According to this view, human nature is sacred, indeed to a lesser extent all nature is sacred, because as creatures we share in God's nature. Through our shared nature, particularly its rational capacities, we are able to learn of the nature of God and come to understand that everything has a complete dependence on God. Thus, the sacred allows us to understand some of God's nature and His actions.

A Contemporary Notion of the Sacred

Contemporary versions of the sacred hold to some aspects of Aquinas's conception without accepting his entire position. One common position claims that the sacred "transcends human affairs in the sense that it is experienced as having value independent of human decisions and preferences." The ethicist Gregory E. Kaebnick wants a source of value rooted in transcendence, analogous to Aquinas's recognition of the meaning of creation. But for Kaebnick, this value-conferring notion of the sacred is free of the historical baggage of God and creation—it has no "special ontological relationship." He suggests (but does not explicitly say) that this notion of the sacred, taken from environmental ethics, is generally about nature, but it can be applied to bioethics issues, including the challenges of genetic engineering. Specifically, Kaebnick contends that the advantage of the notion of the sacred is that it carries with it an acceptance of value that is not defined solely by individual human ends: "The idea of the sacred is the idea that we bear a moral relationship to other things."

Kaebnick suggests that we can "ascrib[e] sacredness to a work of art, the human body, nature, or the natural order of birth and death" as a pre-theoretical response to the world. When he reaches for content for the sacred, he looks to a romantic view of nature characteristic of such poets as William Wordsworth or Walt Whitman, or to the romanticist painters of the Hudson River School. Kaebnick's romanticism involves the wistful hope that we can recall the time when nature and its workings were truly beyond us and, with a proper recollection, we shall once again respond with reverence.

The immediate difficulty with this analysis is that it begs the very question it hopes to address. The era in which we possess remarkable powers to influence or alter biological life is already upon us, and one is hard pressed to find any area of biological life that has *not* been touched by the human hand. Kaebnick's notion of the sacred must thus simply be understood as conservation, a sense that the past was somehow better: "If something's being sacred does not flatly prevent us from acting in ways that appear to threaten that entity, the right thought—drawing again from environmentalism—is that *we ought to tread lightly and cautiously*" (italics original). Yet it is precisely the aggressive trend of technological advance that is at issue.

The key problem in trying to stop—or at least to slow long enough to properly consider—the implementation of such aggressive techniques as genetic engineering by appeal to the sacred is that some understandings of "sacred" are empty. Once the nineteenth-century German philosopher Friedrich Nietzsche famously declared that God is dead, there was good ground to argue that the sacred died with him. For Aquinas, the sacred carried with it the weight of its divine origin. There is a richness of content derived from the rootedness of the sacred in a larger reality, one taken to transcend human reality, and to which humans need access. In considering Aquinas's understanding of nature, one also must consider nature's creator and the creator's divine plan. When God died, the "fact" of that death also removed the possibility of man as an image of God and the animal world as possessing traces of God. Kaebnick's sense of the sacred cannot fill the void left by God's death. Without a source, image and trace evaporate. When they evaporate, so too does our ability to locate those elements in our world, and to extrapolate from them an ethical guide, or even the reason to want to bother to do so.

The Sacred as Feeling

Another notion of the sacred understands it not as a concept, but that which elicits a feeling. One prominent example of this view is medical ethicist Leon Kass's use of the "Yuk factor" as the beginning of an argument against cloning. The "Yuk factor" is an emotional reaction to the violation of an intrinsically valuable limit. The philosopher and ethicist Mary Midgely makes a similar point, insisting that "it is usually a bad idea to see debates . . . as flat conflicts between reason and feeling because usually both thought and feeling are engaged on both sides." She wants us to pay closer attention to feelings as reason's allies and co-directors of ethical action. At the same time, Midgely is also aware that feelings must be articulated by thought and expressed in communication with others. Nevertheless, she calls upon feelings to augment an argument, to provide it with a pre-conceptual justification that eases the way for argument.

The problem with Midgely's attempt to defend emotional revulsion at the thought of genetic engineering—by lauding the usefulness of the emotions as a sort of early warning system—can go only so far. The important question—how the emotions got the form they have that lead them to react as they do—remains unanswered. Further, all horrors are not possessed of the same insight (Midgely mentions a horror of cats, for example) and the differences among emotional responses raise two important questions: 1) What are the social and psychological conditions that give rise to the specific emotional reaction? and, 2) What rational explanation can grow out of that emotional reaction?

Certainly, addressing the social and psychological underpinnings of an emotional reaction is a complicated affair, but Midgely's attempt to draw an argument from emotional reaction yields ambivalent conclusions. She is horrified at two very different—albeit related—issues: first, concern with the genetic alteration of species, and second, concern with the nature and growth of the industry of bioengineering. She quickly moves from a horror at mice growing human ears on their backs to horror at "this huge uncriticized impetus, this indiscriminate, infectious corporate overconfidence, this excessive one way channeling of energy" of the biotechnology industry. She seems to find acceptable "single projects, introduced slowly, tentatively, and critically," but urges delay on the larger, more sweeping projects of biotechnology, insisting they are "unnatural" and "in the quite plain sense [they call] us to alter radically our whole conception of nature."

Although emotion—such as the horror Midgely describes at the prospect of mice with human ears, or fear at the prospect of large genetic projects—offers important *occasions* for reflection, emotion does not *ground* or *inform* reflection. Our emotions also are usually not "pure" responses; they reflect our goals. For instance, one may be repulsed by the thought of a mouse with human ears (in part because one sees no use in a mouse altered in this way) but one might not be horrified at equally strange elements in a research program that would eliminate the possibility of Huntington's chorea or Tay-Sachs from a child's future. Our interests, in short, influence our emotional reactions.

The Hard Case of Genetic Engineering

Despite these difficulties, I want to suggest that the notion of the sacred can help us with the ethical questions raised by genetic engineering. To do this, I turn to Plato, the premier philosopher of the Western tradition. In his analysis of the human psyche and what drives it, Plato suggests that *eros,* a deep psychological force, propels us to go beyond ordinary human needs and capabilities to seek out truth, beauty, and goodness. As Diotima, a main character in his dialogue *Symposium,* says, men are "pregnant in respect to both body and soul," and this pregnancy is a striving for the immortal. This striving takes many forms—from our desire to have children so that we may live on past our death, to contemplation, the reward of the practice of philosophy in which the eternal ideas are thought.

One could suggest that genetic engineering is a concern of those who are pregnant in body only, and thus worries about genetic engineering are "beneath" true intellectual evaluation. When one glimpses immortality through one's children, the issue of genetic engineering seems to confer benefit. Few parents would deny their child any help that would avoid or correct a genetic defect, and likely many parents would also be interested in genetic enhancement.

If Plato is right, then *eros*—that deep psychological force that drives humans to go beyond the ordinary—will resist attempts to preserve the status quo or thwart the promise of genetic engineering to achieve immortality, betterment, perhaps even perfection. Manipulation of the genome promises individuals the power to mold the future according to an image of the beautiful to which they aspire, or perhaps come to expect. Parents will be especially drawn by this promise—anyone who has lost sleep during a pregnancy, fearing the birth of a child because of as yet unknown difficulties, deformity, impairment, death, knows well the implicit ideal that reproduction and birth holds up against an uncertain reality. The relief that accompanies a "normal" birth, or the devastation and regrouping that follows a "problem" birth, give further testimony to power of that ideal. The suggestion that the fate of a child should be left to the hand of cruel and random nature when technical means exist to minimize the possibility of disaster seems ludicrous to the profoundly worried parent.

At this point, the recognition that such technical means are not now fully developed, that they will be risky, complex, expensive, and perhaps unjust, carries little force in the face of the significance of the accomplishment. This is true for those with the resources to gain access to the technology, but it is also true for those who are either passionately worried, risk takers, or are confident—even overconfident—about technological innovation. Further, *eros,* that strong psychological drive to go beyond need and capability, will at times lead one, unknowingly, to risk great harm. The parent who chooses genetic manipulation—out of love and protective feelings for their child—might never consider *other* risks or dangers that the child faces as a result of genetic alteration. Certainly, human history is filled with examples of the terrible consequences to societies that have attempted "improvement" or "perfection" of some of its members, or have deemed some as "better" and others "lesser."

The more one is driven by desire, the less one is willing to consider the consequences of the fulfillment of that desire. If *eros* responds to genetic technology, then it is possible—though not assured—that we might give careful consideration to the implications of alterations to the physical makeup of the human being. Plato addresses this general point as well, describing the human being as struggling with different passions, an appetite for physical pleasure and an appetite for the fruits of reason. The results of this tension are as erratic—and possibly dire—as those of Plato's famously evocative example of the chariot pulled by two temperamentally different horses. A dark horse (the desire for pleasure) is yoked with a white horse (an appetite for recognition under the guidance of either pleasure or reason), with a charioteer (reason) struggling to control them. The progress of this chariot is never assured, as at times the charioteer is in control and at times the dark horse seduces his teammate and they rule the chariot. Humans are often under the influence of their dark horse.

But perhaps we moderns have done a more thorough job of thinking about the dark horse than we give ourselves credit. As Nietzsche insisted, to properly understand nature, we must recognize its overwhelming capriciousness, as the stark and wanton suddenness of disease will attest. Our *eros* (our internal dark horse) often rules us, and—if Nietzsche is correct—it can be encouraged or manipulated. Thus, a simple optimism about our abilities and the comforting embrace of nature are not adequate to our reality.

If we are to generate principles and practices that allow us to reasonably govern the use of genetic technologies, we must find ways to address both our striving for beauty and immortality, and our revulsion at a nature indifferent and wanton beyond measure.

Toward an Answer

Some, such as political theorist Francis Fukuyama, hope to close off the possibility of genetic engineering in order to preserve human nature in its present form. Fukuyama's view of human nature is that which has achieved an end of history in the development of democracy. But he fears that change will begin a new history, and we then must learn to deal with a different human nature. As I suggested, however, the force of *eros* precludes a simple halt to genetic engineering, even in the name of preserving human nature (as we know it). If this is so, then perhaps we can say that Fukuyama is right about the wrong problem. Technical advances in genetic engineering will indeed alter in unknown and unforeseeable ways our very human nature—but only the nature of some. Because of humanity's enormous genetic diversity, for a time the alterations we make will have only a limited effect. Those individuals who are engineered (and their descendants) will comprise only a small proportion of the entirety. Their significance will be magnified by their symbolism of their origin, and to some extent, by the success of the engineer's intent in bringing about the physical improvements promised by genetic engineering. But others will have a different view of perfection, or will not turn to genetic engineering as a reach toward perfection at all. Instead, they might view genetic engineering as a means to restore, sustain, or enhance health in order to continue a life devoted to self-understanding and perhaps to wisdom.

These differences can create stubborn social tensions. But an appeal to the sacred will not establish clear lines between the permissible and the impermissible. Indeed, the attempt might exacerbate the tension. The difficulty lies in trying to use the sacred as a normative category in the absence of a community-shared religious experience. At best, the sacred is a consequence of an already existing community experience. At worst, it will manifest the varieties of incommensurable divine experience, and seemingly justify all manner of genetic experimentation.

Sources

For definitions of 'sacred' see *Oxford English Dictionary* (Oxford University Press, 1971); all references to Thomas Aquinas are from the *Summa Theologica,* found in *Basic Writings of Saint Thomas Aquinas,* edited by Anton C. Pegis (Random House, 1944), vol. 1 (Q1, A5), (Q35, A1), (Q93, A2), (Q93, A6); Gregory E. Kaebnick, "On the Sanctity of Nature," *Hastings Center Report,* vol. 30 (2000); Mary Midgley "Biotechnology and Monstrosity: Why We Should Pay Attention to the Yuk Factor" *Hastings Center Report,* vol. 30 (2000); Plato, *Symposium* (Penguin, 1999); Plato, *Phaedrus* (Penguin, 1973); Francis Fukuyama, *Our Posthuman Future: Consequences of the Biotechnology Revolution* (Farrar, Straus and Giroux, 2002).

Normal Humans, Human Nature, and Genetic Lessons

Robert Wachbroit

Introduction

Genetic discoveries can affect our self-conception. While concerns about what such discoveries will enable us to do—at what cost and with what risk—are of course important, their implications for how we should understand ourselves can sometimes matter more. Whenever scientists identify or even suggest a genetic influence on behavior, we worry that the existence of such influences might be incompatible with our acting or choosing freely. The belief that we have such freedom is central to our common self-understanding. Our understanding of normality—who or what is normal and why—is also central to our self-conception, and it can also be affected by genetic discoveries.

Classifications schemes—whether of genes, behaviors, attributes, or even people—record a variety of differences. But when some differences are then labeled as normal or abnormal, they reveal our special interest in how differences should be interpreted. To put it colloquially, we regard normal differences as OK, abnormal differences as not OK. Our concern with normality is therefore tied to the anxiety—or comfort—we derive from the thought that normality might be objective, reflecting the way the world is and not just convention or human (social) construction. It is easy to believe, or fear, that discoveries in genetics—those fundamental markers of biological differences—have important implications for our understanding of normality.

The matter, as we shall see, is much more complicated. Claims regarding human normality are also thought to have implications regarding human nature. Both human normality and human nature concern differences, but whereas human normality usually concerns the differences between people, human nature usually concerns the differences between people and other things—e.g., other animals. Thus, our understanding of human normality touches on our self-conception as individuals; our understanding of human nature touches on our self-conception as human beings. How these two are connected will occupy us later in this discussion.

The Multiple Meanings of "Normal"

One of the first things we should note about normality is that the word "normal" is used in many senses, which we can group into three broad categories—statistical conceptions, social conceptions, and biological conceptions.

Statistical Conceptions. Examples of meanings or definitions of "normal" that fall into this category include definitions of the normal as the average (or the mean), as the most common (or the mode), or as falling within two standard deviations of the mean. Statistical conceptions of normality are the clearest and most straightforward conceptions of normality. They are clear in their applications—determining the average is a straightforward, mechanical procedure—and also in their limitations. Because they are nothing more than a mathematical property of populations, one cannot take what is normal in one population and generalize or extrapolate that to another population. Knowing the average height in one population, for instance, tells us nothing by itself about the average height in a different or larger population.

Social Conceptions. This category includes a wide variety of conceptions of normality, ranging from ordinary judgments about what is acceptable behavior to more sophisticated accounts of social expectations and deviance. Social conceptions include explicit pronouncements about what is "weird," as well as assessments of social acceptability. Some social conceptions of normality refer to everyday behaviors or conditions; others refer to states that are ideal but unrealizable.

Although statistical conceptions of normality and social conceptions of normality are about different things—and so are logically independent—they may be related by scientific theories or empirical hypotheses, claiming an empirical relationship between the statistically normal and the socially normal. One might hypothesize that, under certain conditions, what is common—regarding sexual behavior, for instance—will be what is socially normal or acceptable. But this is an empirical hypothesis, not a logical or conceptual truth. Specifying the conditions for such interconnections between statistical and social conceptions of normality is, of course, one of the tasks of empirical social science research. Indeed, discoveries that something socially abnormal is nevertheless statistically normal are surprises that can attract popular interest—such as an interest in unmasking widespread hypocrisy.

Biological Conceptions. According to the standard understanding of biology, one of the primary tasks of biological research is the identification and explanation of biological function. It is not enough to know how the organ, tissue, cell, or gene behaves or reacts under various conditions. We need to know its function. What is the *purpose* of this organ? To circulate the blood. What is the purpose of this gene? To produce a protein essential to protect the lungs against airborne irritants. In identifying the biological function of an object, the biologist specifies what it is for that object to be normal or to be in a normal state: it is normal insofar as it is performing its function. Biological differences that do not affect functioning are differences that do not matter to the classification of biological normality. For example, the different blood types do not differ in biological function and so are all biologically normal.

This last example underscores the difference between biological and statistical normality. Although type O-negative blood is rare—that is, statistically abnormal—it is biologically normal. The opposite—statistically normal but biologically abnormal—is also found, as in the case of a widespread epidemic, or dental decay. Even when a disease is statistically normal, it arguably reflects some biological dysfunction—i.e., some biological abnormality—because it is a disease. The biologically normal is also more likely to flourish than the biologically abnormal, and so more likely to be the statistical norm. Even though epidemics can undermine that presumption at a particular time, from the standpoint of the evolutionary history of a species, the biological norm will likely be the statistical norm.

We should also note the difference between biological normality and social normality. For instance, disease is often defined as an abnormality, but no agreement exists about what sort of abnormality it is. The controversy over the definition of "disease" turns on the question of whether disease should be understood as a biological abnormality, a social abnormality, or some combination of the two. The issue would not make sense unless biological and social conceptions of normality were distinct. Further, depending upon the society, a biological abnormality, especially if it is highly visible and statistically abnormal, may also be deemed socially abnormal—it is regarded as a stigma, and the person possessing the biological abnormality is stigmatized. The easy association of biological abnormality with social abnormality can be a matter of some concern.

Genetic Discoveries and Judgments of Normal

Although scientists have yet to discover what constitutes biological normality in many areas, particularly with respect to certain behaviors, we can expect this to change with advances in genetics research. We could well discover that some behaviors fall within the normal range of social behavior even though the behavior reflects a biological abnormality. The possibility of such discoveries in behavioral genetics has important ethical and social implications. In particular, if scientists discover that some behavior is biologically abnormal, there is the danger that it will be relabeled as socially abnormal and result in social stigmatization.

Imagine a person who is somewhat messy—his clothes are a bit disheveled, his workplace looks disorganized, and his home is full of clothes and dishes lying about—but all within the socially normal range. If it were discovered that in some cases such behaviors reflect a biological or genetic abnormality—specifically, a mental abnormality—the result could transform the perception of the behavior from socially normal and accepted to abnormal and stigmatized. (This may seem farfetched at first until we recall the scientific reports not long ago of an alleged "grooming gene.") A similar concern accompanies cases where the social acceptability of a behavior is controversial. Discovery that a behavior is biologically abnormal might persuade some (and give encouragement to those already convinced) that the behavior is indeed socially abnormal. This concern lies behind much of the controversy over the claimed identification of genetic alleles that predisposed to homosexuality: even if sexual orientation was discovered to be involuntary, such claims would likely increase the stigmatization of homosexuals if these alleles were classified as abnormal.

This phenomenon bears some similarity to the process known as "medicalization," where a social type—the elderly, the alcoholic, the drug addict, for instance—is transformed into a medical type and treated accordingly: the problems of the alcoholic are medical problems, solutions are to be found in medical treatment, and the responsibility of the alcoholic is somewhat diminished because of his "medical condition." Medicalization of a social classification is not always wrong; indeed, in some cases, medicalization can arguably be seen as representing moral progress.

The justification for reassessing social classifications based on discoveries in biology will thus depend on the specifics of the case. But reassessing social classifications based on confusions among kinds of normality is always wrong and unjustified. And that is just what is happening in the cases in which a behavior previously considered socially normal is now deemed abnormal simply because of a discovery that it is *biologically* abnormal. Such relabeling rests on a conflation of distinctly different kinds of abnormalities—social and biological. It is simply a mistake to conclude that a behavior is socially discreditable just because of its biological origins.

The adverse social consequences risked by such confusions are part of a larger phenomenon regarding the public reception of scientific discoveries. To what extent do our commonsense beliefs and ordinary practices rest on assumptions of a primitive biology or, as philosopher Bertrand Russell would put it, a "stone-age metaphysics"? We could ask this question about the threat posed by progress in genetics to our central beliefs and practices: Can genetic discoveries undermine our belief in free will and ordinary practices regarding moral responsibility? Can they challenge our ordinary beliefs regarding what is a fair and just distribution of resources? Can such discoveries undermine our ordinary beliefs about who is sick and who is healthy?

Certainly, we need to educate the public and, where appropriate, scientists, about these distinctions regarding normality and how to avoid confusing one type of normality with another. Nevertheless, not all reassessments of social normality based on biological discoveries involve crude conflations of different types of normality. One type of sophisticated response in particular warrants special attention. This kind of response to genetic discovery does not so much turn on the idea of normality as on the idea of human nature, because discoveries about what constitutes (biological) normality are seen to have implications about what constitutes human nature. With a proper understanding of human nature, we are led not only to better self-understanding but also to better public policy.

The Normal and the Natural

A good illustration of this type of sophisticated response can be found in Francis Fukuyama's recent book, *Our Posthuman Future.* Fukuyama holds that the relationship between the normal and the natural—specifically, between human normality and human nature—is roughly that of identity.

> The definition of the term *human nature* I will use here is the following: human nature is the sum of the behavior and characteristics that are typical [i.e., statistically normal] of the human species, arising from genetic rather than environmental factors.

With this definition Fukuyama argues against biotechnological modifications that go against human normality and so against human nature. Such modifications would violate the border that separates humans from other animals and would be morally objectionable, according to Fukuyama, because our dignity as human beings is based on what distinguishes us from other animals—i.e., our human nature. Our sense that we owe a greater moral obligation to human beings than to other animals is based on our shared human nature. Our moral community would be destroyed if we allowed genetic interventions on human beings that resulted in creatures that lacked this nature. Since human nature is our "genetic endowment," scientific discoveries detailing this endowment, specifying human normality, should help identify what genetic interventions are morally permissible.

With this much allegedly at stake, we should look more carefully at Fukuyama's definition of human nature, which he intends to be statistical: "typicality is a statistical artifact—it refers to something close to the median of a distribution of behavior or characteristics." Fukuyama invokes a statistical concept in his definition of human nature in order to allow for exceptions or mutations. (Fukuyama gives the example of the mutant female kangaroo born without a pouch, which does not undermine the claim that having a pouch is part of (female) kangaroo nature. Although he is not explicit about this, I assume that Fukuyama holds that a pouch-less female kangaroo is an abnormality and not a mere variant of the species.)

Insofar as Fukuyama wants his definition of human nature to identify what is distinctive of the species *Homo sapiens,* his appeal to statistics is misplaced. If the concept of human nature refers to the identifying features of the species, then there is no need to make any mention of what is typical or statistically normal in the population.

Species can undergo hard times, with most of the surviving members suffering disease or trauma, so that the average member of the species might not be representative of the species. If most current kangaroos succumbed to a disease, leaving mainly the pouch-less kangaroos, we would not conclude that pouches on females were no longer distinctive of the kangaroo species and that pouches are not part of kangaroo nature. The conclusion would conflate statistical normality with biological normality. Instead we would say that kangaroos were on the verge of extinction, with primarily abnormals or mutants surviving at the moment, or we might say that a new species, similar to *Macropus rufus* (i.e., kangaroo) was emerging. Fukuyama would do better to employ the concept of biological normality rather than statistical normality in his definition of human nature.

Perhaps a more troubling feature of Fukuyama's definition of human nature is the final phrase, "arising from genetic rather than environmental factors." On the literal reading of that phrase, we are to consider those behaviors or characteristics that are caused by genes only, in which environmental factors play no causal role. The difficulty with this reading is that it conflicts with a basic principle of genetics: the characteristics of an organism—its phenotype—are the result of a causal interaction between the genes of the organism—its genotype—and its environment. There is no characteristic or behavior that is a product only of genes. Nor is the environment some unimportant or constant background condition.

It may be that what Fukuyama really meant to write in that last phrase was "arising *more* from genetic than environmental factors." But now two difficulties arise. First of all, it is not at all clear what "more" means in this context. In physics we can easily determine whether one cause is more significant than others and by how much, using such standard techniques as the vector decomposition of forces. But no such technique is available in genetics, in part because the impact of genes is not independent of the environment: genes are one factor, the environment is another factor, and, in an important sense, the interaction between genes and environment is yet another factor. The second difficulty lies in the apparent arbitrariness of "more." Are we to believe that a characteristic that is 51 percent genetic is part of human nature but a characteristic that is only 49 percent genetic is not? Because of this confusion and obscurity in Fukuyama's definition of human nature, it is hard to see how that concept could guide or even constrain, public policy.

Leaving Fukuyama's troubled definition, let us turn to a much different understanding of the relationship between human nature and human normality. According to philosopher of science Ian Hacking,

> "Normal" bears the stamp of the nineteenth century and its conception of progress, just as "human nature" is engraved with the hallmark of the Enlightenment. We no longer ask, in all seriousness, what is human nature? Instead we talk about normal people. We ask, is this behaviour normal? Is it normal for an eight-year old girl to . . . ? Research foundations are awash with funds for finding out what is normal. Rare is the patron who wants someone to investigate human nature.

In contrast to Fukuyama, Hacking suggests not that human normality and human nature are identical but, instead, that the scientific concept, human nature, has been shown to be wanting and has been replaced by the scientific concept, human normality.

While it is certainly true that the term "human nature" is rarely used in the current scientific literature, the term has not gone the way of other discarded scientific concepts such as phlogiston or the ether; it is regularly invoked in the humanities and in the popularization of science. This suggests that the relationship between human normality and human nature is more complex than Hacking contends.

One element of this complexity is that the effort to characterize human nature is often the attempt to formulate the scientific laws and principles governing human behavior, emotions, cognition, etc. (For some, particularly those in the grip of the so-called "nature vs. nurture" debate, the crucial issue is whether these laws can be specified independently of the environment.) The concept of human normality seems well suited to investigate many of these issues, transforming these questions to ones presumably more amenable to scientific investigation—e.g., what constitutes (biologically) normal human behavior?

In other instances, the concept of human nature is invoked to refer to what is distinctive about human beings. What makes something a human being—rather than a chimpanzee, a tree, or a rock—is that it possesses a human nature. Simply identifying the laws governing human beings does not address this issue of distinctiveness unless these laws are shown to be *necessarily unique* to human beings.

The concept of human normality is not well suited to address these issues. If human nature refers to the essential property of being human, then a human being who did not possess human nature would be equivalent to a human being who was not a member of the human species—which is a contradiction. Biological normality, however, does not constitute a criterion of identity. An abnormal or malfunctioning heart is still a heart, not a different organ. To be sure, we could imagine altering the heart so that it became a different thing altogether, but then we would not call the result an abnormal heart: a heart that was burnt to ashes would no longer be a heart; it would not be an abnormal heart. To put the point more generally, a human being who was not normal would simply be an abnormal human being—but a human being nonetheless. This is why investigations regarding human normality are only aimed at explaining differences among human beings. Discoveries about the differences among humans do not entail any conclusions regarding the differences between humans and other beings.

Investigations in human genetics can help us better understand human (biological) normality. It can help us determine whether a particular difference between humans is a mere variant or an abnormality. But an investigation in human genetics—such as the human genome project—cannot determine what is essential to being human. For that we need at least the findings of various ape and chimpanzee genome projects. (We should note that, for some commentators, even the findings of these other projects would not be enough for identifying what is essential to being human because the structure of evolutionary theory complicates the question of what are the essential properties of being human.)

These considerations are bad news for those like Fukuyama who want to draw large claims about human distinctiveness from genetic discoveries. Current investigations in human genetics, including those regarding normality, seem incapable of identifying the essential properties of being human. But even if such properties were discovered, their discovery would not entail social or moral conclusions. We cannot go from a view of what is essential or distinctive about humans to a view of what makes humans special. That would be like saying: Having an atomic number of 79 is what's essential to gold, and so that is what makes gold (financially and esthetically) special. Such a conflation is comparable to the conflation of biological and social normality. Discoveries in human genetics will reveal much about the biologically normal; but these discoveries should not be taken as conclusions regarding what should be socially normal, or what is essential in human beings.

This article derives from "Normality and the Significance of Difference," which will appear in *Wrestling with Behavioral Genetics: Implications for Understanding Selves and Society,* edited by Erik Parens, Audrey Chapman, and Nancy Press (Johns Hopkins University Press, forthcoming).

Sources

One example of a sophisticated account of social expectations and deviance is Erving Goffman's classic study, *Stigma* (Prentice-Hall, 1963); on the "grooming gene," see Joy Greer and Marion Capecchi, "Hoxb8 Is Required for Normal Grooming Behavior in Mice," *Neuron,* vol. 33 (2002); Francis Fukuyama, *Our Posthuman Future: Consequences of the Biotechnology Revolution* (Farrar, Straus and Giroux, 2002); Ian Hacking, *The Taming of Chance* (Cambridge, 1990); for a fuller discussion of human essentialism, see E. Sober, "Evolution, Population Thinking, and Essentialism," *Philosophy of Science,* vol. 47 (1980).

The Ethics (and Politics) of Genetic Technologies

III

Finessing Nature

Richard M. Zaner

Introduction

Even those who favor the prospect of human cloning have become troubled, largely because, they believe, so many of us in this complex society are simply too *laissez faire,* too jaded to appreciate the altogether serious implications of scientific endeavors. We sometimes seek too avidly advances in biomedical science that are meant to benefit us and are rich with promise for our future children and for humanity—but these advances also can risk the very things we cherish most. This applies especially to the prospect of human cloning. Declared over four decades ago, the sentiment expressed by Nobel Laureate geneticist Joshua Lederberg in 1966 is captivating still: "cloning could help us overcome the unpredictable variety that still rules human reproduction, and allow us to benefit from perpetuating superior genetic endowments."

That theme was later adopted by other Nobel Laureates in genetics—Sir John Eccles, Sir Macfarlane Burnett, and more recently by Walter Gilbert and James Watson. Scientists seem a united front. Their vision embodies the notion that it is best that "nature" *not* be left to its own devices, as it is essentially blind and unpredictable. Therefore, nature must be, as I prefer to say, finessed. Human beings must actively support scientific efforts to take matters into their hands through planned interventions. Those who oppose genetic manipulation and cloning are not only wrongly suspicious of science, proponents argue, but also irrationalism and numeric illiteracy also often account for their suspicion.

This orientation toward human life and reproduction outrages those most opposed to cloning and manipulation of the human genome. Expressed energetically some decades ago by biomedical ethicists June Goodfield and Paul Ramsey, most recently this view has seen vigorous expression by Leon Kass, chair of President George W. Bush's Council on Bioethics. (The Council was established on October 3, 2001, following the expiration of the charter of the National Bioethics Advisory Commission, NBAC, which itself was created by Presidential Executive Order in 1995, during the Clinton administration.) Kass had long been a critic of the NBAC, worrying that it would yield "to the wishes of the scientists to clone human embryos," and would call for only a temporary moratorium on "implanting cloned embryos to make a child." The NBAC did indeed recommend only a temporary ban. Kass and other scientists, who are deeply suspicious of genetic manipulation generally, and of cloning especially, embrace the belief that the so-called "natural" way is indeed the preferable way, because it is ultimately ordained by God. (Further complicating matters is that, while opponents of cloning hold views that typically rely on some understanding of religion, they rarely seem to find a place for religious language in their defense of cloning, stem cell research, and genetic engineering.)

The purpose of this article is to suggest that we need to consider carefully the ethical implications of substituting technology and genetically innovative means to assist human reproduction—that is, of "finessing" nature. To do this, one must first understand some of the ethical debates surrounding the most contentious form of assisted reproduction that lies in the future—that of human cloning. After examining the arguments against cloning, I present thoughtful reconsiderations of those arguments (some of reconsiderations actually support the prospect of human cloning). Finally, I discuss the work of two Canadians researchers, who suggest that environmental conditions might actually accelerate the use of reproductive technologies. I conclude by proposing the outlines of the ethical stance we must take in order to approach with care the prospect of finessing nature.

Arguments against Cloning

1. Cloning is repugnant. Kass's appointment as chair of the President's Council on Bioethics suggests that a new day has dawned for the opponents of cloning and related research. The trouble with human cloning is that it is fundamentally "dehumanizing" and everyone ought to find it repugnant. At the crux of his argument, oddly, is the simple and undefended assertion that cloning should evoke a common feeling—"repugnance"—that points to a "deep wisdom, beyond reason's power fully to articulate." Even Ian Wilmut apparently agrees: during his initial announcement of Dolly, the first cloned mammal, he declared that using his technique to produce human beings would be "quite inhuman."

2. The "natural" way is the "profound" way. Kass also argues that human beings are sexual in a special sense: reproduction in its "natural" form brings about "some very special and related and complementary" relationships to others. The power of sexuality, he finds, "is, at bottom, rooted in its strange connection to mortality, which it simultaneously accepts and tries to overcome." In contrast, human cloning is essentially asexual reproduction; the organism that is divided to become two is itself preserved (doubly), and nothing dies. Sexuality, however, "means perishability and serves replacement; the two that come together to generate one soon will die. . . . Whether we know it or not, when we are sexually active we are voting with our genitalia for our own demise. . . . " Therefore, he continues, "sex is bound up with death, to which it holds a partial answer in procreation."

This understanding of human sexuality allows Kass to set forward his main objections to cloning. He believes that cloning creates serious issues of identity and individuality. Not only will the clone be genetically identical to another human being, but the clone also might be twin to the person who is his "father" or "mother," adding "psychic burdens" to being the child or parent of one's own twin. The clone will also "be saddled with a genotype [genetic constitution] that has already been lived. He will not fully be "a surprise to the world," but will instead be used goods. Kass also argues that cloning is a giant step toward making procreation manufacture—"a process already begun with *in vitro* fertilization and genetic testing of embryos." Cloning will allow "human artisans" to select "the total genetic blueprint of the cloned individual." In this kind of reproduction, the maker is not equal but superior to the one made, who is a product designed to "serve rational human purposes." Cloning, finally, Kass insists, confuses and disrupts the human relationship, introducing a "profound and mischievous misunderstanding of the meaning of having children and of the parent-child relationship." Natural procreation *means* that by their act the partners affirm the creating of a new life, blending themselves into a baby with its "own and never-before-enacted life to live."

3. The "bright line": here and no further. Concerned that the NBAC did not represent "the voices of grassroots Americans," a group calling itself The American Life League, Inc. established in 1997 the American Bioethics Advisory Commission (ABAC), which subsequently issued its own report. The ABAC stressed that there exists a "bright line dividing legitimate and exciting scientific research and development from dangerous and dehumanizing manipulation." The only way to "embrace what is good and reject what is evil," is to respect this line by encouraging "responsible experiments in animal cloning," but imposing a global ban on human cloning. The mention of "good" and "evil" were meant to raise religious concerns that, according to ABAC, were absent in debates of this kind.

4. You just can't trust 'em. In fact, the ABAC also contended that the rare mention of "religion" is often accompanied by embarrassment. The NBAC apparently agreed with this point and, enlisting the assistance of several groups, authorized a study of the religious issues and themes raised by human cloning. One group, the Program for Ethics, Science, and the Environment (housed at Oregon State University's Philosophy Department) sought the view of persons of faith from a wide variety of traditions. Worries about the prospect of human cloning varied, with, for instance, one clergyman speculating that "cloning would violate practically every sacramental dimension of marriage, family life, physical and spiritual nurture, and the integrity and dignity of the human person." But another clergyman suggested that cloning should be permitted to continue, but "the impetus and ethical motivations behind human cloning [must] be meticulously monitored," its practices "tightly regulated," cloning of human cells must never be permitted to reflect "racial or ethnic, or other demographic subgroup (i.e., gender)" and, finally, access to any benefit must be universal.

The range of recommendations by religious thinkers reveals that technology is not itself disturbing; rather, these thinkers seem to worry that scientists and physicians, in whose hands genetic manipulation is entrusted, are more likely to break than to keep that trust.

5. The reach of myth. Wendy Doniger, a distinguished history of religions scholar at the University of Chicago, suggests that the notions of threat and promise that accompany the prospect of cloning human beings are deeply embedded in Western history and mythology. Although the actual prospect of cloning a human being is quite recent, Doniger points out that something very much like it has fired the human imagination for millennia and finds expression in a variety of myths. She is particularly interested in the historically embedded notion that "doubles could be produced by the ancient counterpart of science—magic."

According to her research, a main reason to oppose the making of "doubles"—combined with fear and fascination, not unlike today—had to do with attitudes about sexual conduct: "time and again one clone somehow or other stumbles in the other's bed, and this feel of advertent or inadvertent sexual betrayal" is "an inescapable part of the terror of cloning." Mythic sources suggest that clones—or "doubles," "duplicates," "multiples," erase the individuality of the one doubled and thus threaten personal identity. Another main mythological theme concerns eugenics, the search for ways to use biological means to "improve" future humans. The sexual side predominates here as well, since the principal theme is that of males who want to ensure that their offspring look like them—guaranteeing their paternity as well as the presumed quality of their offspring. At the heart of these myths is an elemental resistance to the possibility "that a clone—that is, a magically created double—could have a separate soul, that in creating a body *de novo*, the magician could create a soul *de novo* too."

Responses to the Objections

Although much has been made of the presumed repugnance at the prospect of human cloning, the subversion of the natural, and the inappropriate power of its practitioners, challenges to the arguments against cloning are put forward just as vigorously.

1. Questioning the "identical" nature of clones. The well-known Harvard University scientist Stephen Jay Gould has noted with palpable irony how fashion often governs not just public mood but even scientific projects. Much of the knowledge and technical skill needed to clone humans has been around for some time now, and Wilmut's Dolly was not in fact the first mammalian clone—it was merely the first engineered from an adult cell. Gould asks the obvious question: what's the big deal anyway? To which his answer is blunt and immediate: no big deal at all, *if and when* you think about it.

The trouble is, not many have thought about cloning. Not only had Wilmut and his colleagues also cloned sheep from the cells of a nine-day embryo and a twenty-six-day fetus (with much greater success, incidentally), but far and away the most significant of the seemingly endless outpouring of questions about human cloning have already been answered, in good old empirical fashion: "We have known human clones from the dawn of our consciousness. We call them identical twins—and they are far better clones than Dolly and her mother." Gould thus wonders, "Why have we overlooked this central principle in our fears about Dolly? Identical twins provide sturdy proof that inevitable differences of nurture guarantee the individuality and personhood of each human clone."

The preeminent twin researcher, Nancy L. Segal, emphasizes just this point, insisting that "*identical twins are clones, but clones are not identical twins.*" Cloned human beings are not identical in the strict sense "because they fail to fulfill the three twinship criteria: simultaneous conception, shared prenatal environments and common birth." Moreover, cloned individuals differ in other significant ways, because environmental factors will vary; thus, she stresses, however one feels about the perennial dispute, "the *nurture* part of the nature versus nurture equation will be *completely* different." Finally, as many researchers and medical ethicists have noted, not even identical twins are, strictly, *identical:* differences in fingerprints, organic brain structures, intelligence, personalities, and the like are the rule, not the exception. Thus, even if clones were more like naturally occurring identical twins than they in fact are, one would still find significant differences, the sum of which demonstrate the complete falsity of the notion that a clone is an exact copy of a person.

2. Reconsidering the "Yuk factor." Evolutionary biologist Richard Dawkins clearly agrees with the considerations of Gould and others, to which he adds that he finds nothing wrong with human cloning. To those, like Kass, who appeal to "repugnance," Dawkins insists that this "'yuk reaction' to everything 'unnatural'" should make the rest of us stop and think about such a "reflex and unthinking antipathy." In fact, anyone disturbed by scientific interventions into the "natural" order of things should consider the fact that, for instance, it is just as *unnatural* to read books, drive an automobile, cut up carrots, or wear clothing, as it would be to clone babies. Correlatively, it is just as *natural* to clone an infant in order to acquire more healthy cells as it is to use antibiotics to help heal wounds.

As Gould and Dawkins make clear, those who rely on the ambiguous distinction between "natural" and "unnatural" in their opposition to human cloning must accept the onus of showing that it is "any more momentous than the introduction of antibiotics, vaccination, or efficient agriculture, or than the abolition of slavery."

3. We are not in our genes. George Johnson, a science correspondent for the *New York Times,* nicely summarizes several key points in the dispute, observing that "the queasiness many people feel over the news that a scientist in Scotland has made a carbon copy of a sheep comes down to this: If a cell can be taken from a human being and used to create a genetically identical double, then any of us could lose our uniqueness. One would no longer be a self." The anxious sense that human cloning poses challenges to personal identity has deep mythic and biological roots; but here too, when one thinks about it, there is little to fear. Johnson's rendition of the point is worth citing at length:

> That each creature from microbe to man is unique in all the world is amazing when you consider that every life-form is assembled from the same identical building blocks. Every electron in the universe is indistinguishable . . . all protons and all neutrons are also precisely the same. . . .
>
> Every carbon atom and every hydrogen atom is identical. When atoms are strung together into complex wholes—the enzymes and other proteins—this uniformity begins to break down. Minor variations occur. But it is only at the next step up the ladder that something strange and wonderful happens. There are so many ways molecules can be combined into the complex little machines called cells that no two of them can be exactly alike.
>
> Even cloned cells, with identical sets of genes, vary somewhat in shape or coloration. . . . But when cells are combined to form organisms, the differences become overwhelming. A threshold is crossed and individuality is born.
>
> Two genetically identical twins inside a womb will unfold in slightly different ways. The shape of the kidneys or the curve of the skull won't be quite the same. The differences are small enough that an organ from one twin can probably be transplanted into the other. But with the organs called brains the differences become profound.

Although one can argue with the conclusion Johnson draws: "While it is possible to clone a body, it is impossible to clone a brain," he is correct that from the earliest beginnings to the most sophisticated forms of human life, our human genes have the capability of indicating only the *rough outlines* of neural wiring. Random mutation and environment influence neural connections and all aspects of the organism.

4. "Success" in perspective. In a published adaptation of a lecture on cloning delivered at a joint meeting of the medical societies of San Bernardino and Riverside Counties (California), Mark D. Eibert makes a case for encouraging human cloning. Eibert points out that in the case of Dolly, Ian Wilmut began with 277 reconstructed eggs (eggs that had their nuclei removed and were then fused with an adult cell). These eggs were then cultured in sheep oviducts; of the original 277 reconstructed eggs, only 29 successfully divided and became embryos. These 29 were transferred to the uteruses of 13 sheep (some received one, others two or three eggs), but only one sheep became pregnant—from which Dolly was born.

Eibert emphasizes that, comparing this figure—one success from thirteen attempts—to the ratio of live births resulting from *in vitro* fertilization (IVF), it was not until 1990—twelve years after the birth of the first IVF baby in England—that the average success rate for IVF "got to be *as good as* one out of 13." (Currently, the IVF success rate worldwide is about 25 percent, a rate that took twenty years of highly focused efforts.)

But Dolly was only the first cloning experiment of its kind. In the second—cloning of fifty mice in Hawaii—the efficiency rate was even better (again, comparing the number of eggs per live birth): ten times higher than Wilmut's experiment. The third published adult cell cloning experiment, Eibert points out—this occasion, the cloning of cows in Japan—was seventeen times more efficient than Wilmut's results. Moreover, a variety of other species have already been cloned using Wilmut's technique—goats, pigs, rhesus monkeys—"literally hundreds of animals in the world who were conceived through adult cell cloning." Because none of these efforts have been nearly as risky as the detractors constantly allege, Eibert concludes that the argument against cloning based on safety concerns is getting weaker almost daily.

Indeed, for those people unable to have children (between 10 to 15 percent of the population) and for whom IVF or alternative means have failed, cloning might be the only chance to have children. Eibert notes that arguments used more than twenty years ago in the effort to stop IVF are now being used again, this time against human cloning. Before the birth of Louise Brown, the first baby born from IVF, the large majority of Americans polled (85 percent) were opposed to IVF—until successes began to receive public notice, parents of IVF children became more widely known and more commonly encountered, and "safety" a demonstrably irrelevant issue. Precisely the same, Eibert believes, will inevitably occur when human cloning is finally with us, as it most assuredly will be. Human cloning, just as IVF, will become an accepted way of finessing nature's shortcomings.

New Considerations

A quite different and fascinating case raises important questions about human cloning and other techniques that scientists and physicians use to finesse the natural processes involved in human reproduction. Two Canadian sociologists—Louise Vandelac and her research assistant Marie-Hélène Bacon of the University of Quebec at Montréal—argue that "paradoxically" the significant worldwide increase of environmental pollutants, which have been shown to disrupt the human endocrine system, "may accelerate the use of reproductive technologies . . . and even cloning, as well as the dissemination of genetically modified organisms."

Breast cancer, endometriosis, the decline in animal and human sperm density and potency over the past fifty years, as well as a number of maladies have been directly associated with the sharp increase in pesticide use and environmental organochlorine chemicals such as polychlorinated biphenyls (PCBs) and hexachlorobenzene (HCB). Vandelac and Bacon argue that there arise serious, long-term *ethical* consequences of the choices now faced by all nations, individually and collectively. On the one hand, "we could adopt the principles of caution, solidarity and concern for public health," which lead to fewer industrial pollutants, more environmental sensitivity, as well as worldwide education and sound policies that can prevent future damage—a direction they recognize is unlikely to be followed, given the values of a world market economy. On the other hand is the current course of action, in which existing rates and kinds of environmental degradation are permitted to continue and even increase, elevating, in Vandelac and Bacon's words, "the technologization of conception and the 'geneticization' of the living." Since the first option is unlikely to see adoption, continuing the current trend could yield, they speculate, to:

> . . . the growth and widespread institution of a merchant or institutional economy of artificial reproduction, which would gradually replace natural conception for wealthy populations in the Northern hemisphere . . . . We may also witness the proliferation of research

> on embryos, soon to be cloned, patented, genetically altered, or even used as a vector of cellular lines, thus giving the term 'producer' a whole new meaning. This could also lead to . . . the alteration of the concept we have of human beings and humanity itself, an alteration which has already begun due to the technologization of conception, will be further amplified, giving an unprecedented depth to the mutation of human species.

This kind of future would not necessarily result from the rise in endocrine disruptors alone, but this increase in genetic manipulation and resultant compromise to human reproductivity "could be simply used as an easy justification" for "technologization" and "geneticization," with human cloning waiting in the wings for the call to move to center stage. This process is similar to *in vitro* fertilization, which made its way into the center of the reproduction industry. The expected decline in traditional reproductive ability, Vandelac and Bacon content, might lead us to finesse nature, substituting technologically and genetically innovative means to reproduce.

Their conclusion is interesting: "if" the market is allowed to move ever further into human reproduction; "if" the boundaries between and among species continue to be eroded by innovations in gene splicing; and "if" by widespread indifference and apathy humanity is allowed redefinition and manipulation, then, they speculate, we shall have been witness to, and even participants by default in, our own degradation if not demise—one as definitive and final as the disappearance of the dinosaurs. While this is a lot of freight to pile on such an 'iffy' platform, their point is intriguing and merits careful deliberation, for it lays out with clarity an entire series of hard choices we must face squarely and at some point decide.

Toward a Conclusion

To consider these matters clearly, we must conceive of, in the words of Vandelac and Bacon, "a global ethics of the living," embracing environmental concerns as requisite counterpoint to uncontrolled growth of technoscience. We must also make certain to "impose a new ethics of politics," and, finally, ethics must be conceived as perhaps humanity's sole remaining way to counteract the market pressures to "redesign the human . . . ," working toward "a new ethics of economy."

One need not wholly agree with their analysis and plea, but it seems to me that only if ethics is reconceived along these lines is there even the slightest change that it can gain a seat at the corporate conference tables and the halls of government. Only such a reconceived ethics can take an effective part in the much-needed broad and democratic ethical debate that can alone assess, Vandelac and Bacon argue, "this future that some want to impose on others." As the eminent ethicist Hans Jonas once remarked, poignantly if also with some sense of lament, we need wisdom most when we hardly believe any longer in its very possibility—yet need it we do, and if we fail to rise to the monumental task of a new ethics capable of addressing itself directly to "exorbitant power," we fail ultimately in the most solemn way our very humanity.

I close with the razor-sharp words of a poet:

How then shall we make love?
The alligators, with their tiny testicles, mourn in the evergladed swamps
And frogs hop on several of their many legs in Minnesota,
While women on Staten Island hug their barren chests and yearn for breasts

What? We did this to ourselves?
Diminished the very thing that kept us fertile?
Laughed at the gods that gave us better living through chemistry, and
Then tried to sell us the cure for their product-aided ailments.

If there be butterflies still, then there be life . . .
If there be moths, and bees and birds, then there be life . . .
Somewhere, in test-tubes, frozen, gleaming sperm-like histories
Will be what remains of our love-making —
The tiny Jacks and Jills that have survived to tell the story,
To sing our love-sick-song.

—June Zaner, July 2002

Sources

Joshua Lederberg's quote appears in an article published in the *Washington Post* and is cited by Leon R. Kass, "The Wisdom of Repugnance," *New Republic* (June 2, 1997), and Kass's article was later published along with a supposed rebuttal article by James Q. Wilson, in the volume, *The Ethics of Human Cloning* (The American Enterprise Institute Press, 1998); Ian Wilmut's "quite inhuman" remark is found at www.cnn.com (March 12, 1997); Leon Kass's three main objections to cloning are reiterations of objections made by Hans Jonas, in his *Philosophical Essays: From Ancient Creed to Technological Man* (Prentice-Hall, 1974) and also Paul Ramsey's "Shall We 'Reproduce'? II. Rejoinders and Future Forecast," *Journal of the American Medical Association*, vol. 220 (1972); The "voices of grassroots Americans" description occurs in: *Letter to President William J. Clinton from the American Bioethics Advisory Commission* (a project of American Life League, Inc., Stafford, VA), June 21, 1997; "Ban Human Cloning," *Report of the American Bioethics Advisory Commission*, unpublished, a Project of American Life League, Inc., 1179 Courthouse Rd., Stafford, VA 22554 (the Report was sent to President Clinton on June 21, 1997); Wendy Doniger, *The Bedtrick: Tales of Sex and Masquerade* (University of Chicago Press, 2000); Stephen Jay Gould, "Dolly's Fashion and Louis's Passion," in *Clones and Clones: Facts and Fantasies About Human Cloning*, edited by Martha C. Nussbaum and Cass R. Sunstein (W. W. Norton & Co., 1998). Gould, who, at the time of his death, was Alexander Agassiz Professor of Zoology and Professor of Geology at Harvard University and was also widely acknowledged as a preeminent theorist of evolution, offers as proof that "differences of nurture guarantee the individuality and personhood of each human clone" by citing the failures to clone Hitler in *The Boys from Brazil*, and the remarkable individuality evidenced in the famous case of Eng and Chang, the original Siamese twins: one was "a morose alcoholic, the other remained a benign and cheerful man." Nancy L. Segal, *Entwined Lives: Twins and What They Tell Us About Human Behavior* (Penguin, 2000, and republished from the 1999 Dutton edition); Richard Dawkins, "What's Wrong with Cloning?" appears in the Nussbaum and Sunstein anthology, as does George Johnson, "Soul Searching;" all quotes from Mark D. Eibert, Esq. occur in "Human Cloning: Myths, Medical Benefits and Constitutional Rights," *U&I Magazine*, and available from the Human Cloning Foundation at http://www.humancloning.org/users/infertil/humancloning/htm; for a discussion of IVF success rates and how they are determined, see the Center for Disease Control's site at: http://www.cdc.gov/nccd-php/drh/ART00/PDF's/ART2000.pdf; Louise Vandelac and Marie-Hélène Bacon, "Will We Be Taught Ethics by Our Clones? The Mutations of the Living, From Endocrine Disruptors to Genetics," *Baillière's Clinical Obstetrics and Gynaecology*, vol. 13 (1999).

Stem Cell Research and the Legacy of Abortion

Robert Wachbroit

Introduction

On August 9, 2001, President George W. Bush issued a statement in response to a growing controversy that recalled the passionate and bitter debates over abortion. Although the controversy and the statement were not about abortion but the seemingly esoteric topic of stem cell research, critics raised concerns familiar from the debate over the ethics of abortion. Stem cell research, they alleged, is yet another assault on the "unborn." It undermines human dignity and cheapens the value of human life. But advocates of the research were quick to reject the ready linkage to the abortion controversy and claimed that the tools provided by stem cells are among the most important since the development of recombinant DNA techniques in the 1970s. These tools would surely help researchers make discoveries of enormous medical benefit to everyone.

The president's decision—an effort to split the difference by allowing the research to continue on the sixty or so stem cell lines that he believed already existed—satisfied no one but his fiercest loyalists. This essay examines the ethics of stem cell research and some of the difficulties over framing a compromise.

Some Background

It is necessary to begin with a brief description of what stem cells are, why they are of such biomedical interest, and what major ethical issues are raised by the proposed research methods.

Nearly every cell in the body contains the same set of genes (the only exceptions are the cells constituting sperm or unfertilized eggs, which have only half the set of genes found in the other cells). The genes that fix many of the characteristics of the hair are not only located in the hair cells but also in nearly every other cell of the body, including the cells constituting the heart. Nevertheless, the heart does not normally grow hair because not all of the genes in any cell are active; although the heart cells contain the genes for hair, these genes are turned off. Cells, such as those that constitute the heart or the hair, are completely differentiated.

Not every cell is a differentiated cell. Undifferentiated cells fall into one of three types—multipotent, pluripotent, and totipotent. A multipotent cell is an undifferentiated cell that can become or grow into any differentiated cell of a particular type of tissue. For example, hematopoietic cells are multipotent cells that can become or grow into any of the blood cells, such as red cells or white cells, depending upon the body's need. A pluripotent cell is an undifferentiated cell that can become or grow into *any* differentiated cell of any tissue. It has the potential to turn into a heart cell, a hair cell, a blood cell, etc. A totipotent cell is an undifferentiated cell that can grow not merely into this or that tissue but into an entire organism. These three types of undifferentiated cells form a progression: totipotent cells can generate pluripotent cells, which in turn can generate multipotent cells, and these in turn can generate the many differentiated cells constituting the body. These three types of undifferentiated cells are called "stem cells."

Multipotent stem cells should be present in the human body insofar as the body regularly needs to maintain and replace the corresponding differentiated cells. While scientists have located some kinds of multipotent cells, such as those associated with blood tissue (the hematopoietic cells) or those associated with bone and connective tissue (the stromal cells), others, such as those associated with neural tissue, have so far not been identified. Nevertheless, scientists believe that multipotent stem cells of all kinds are present and needed in the body throughout its life.

Pluripotent stem cells, in contrast, are present only during approximately the second week of human development, when the entity is still an early embryo. Soon after fertilization, the fertilized egg or zygote undergoes cell division. By the fourth or fifth day, the embryo has become a loose collection of sixteen identical cells, at which point it develops into a more cohesive cellular aggregate—called a blastocyst—with a central cavity and an inner cell mass. The cells constituting this inner mass are pluripotent stem cells. If these cells are extracted, the blastocyst, and so the embryo, is destroyed.

As this progression suggests, totipotent stem cells are present only during the first few days after fertilization, until the embryo develops into a blastocyst. It is theoretically possible, though technically challenging, to extract one of these cells from the collection of four, eight, or sixteen cells that have developed without harming the others. Not only can the remaining cells develop into a blastocyst and then into a later stage of the embryo, but the extracted cell can also be induced to divide a few times and itself develop into a blastocyst, etc.

The primary interest in stem cell research is with the pluripotent cells. It is not difficult to see why. These cells could be extremely useful in the treatment of any disease arising from cell death or dysfunction, such as Parkinson's disease, diabetes, or certain forms of heart disease. They might, for example, be used in a procedure to replace the damaged neural cells in Parkinson's or the damaged pancreatic cells in diabetes. But the interest in stem cells is not confined to the clinic. Studying stem cells may help us understand the great mystery of developmental biology—how some genes in a typical cell get turned on (i.e., express themselves) while others do not. Despite these promising prospects for theoretical understanding and clinical intervention, a fierce controversy over the ethics of engaging in stem cell research has arisen. It begins with the realization that obtaining pluripotent stem cells involves destroying the embryo.

Not Abortion

The concern over destroying an embryo can easily suggest that we are dealing with a variation of the abortion controversy. Nevertheless, the ethical debate over stem cell research is different from the debate over abortion, and it is important to see why.

The abortion debate is typically framed by two propositions: 1) the conflict is between a woman's right to control her own body, particularly her pregnancy, and the fetus's (or in this case, the embryo's) right to life; and 2) abortion is seen as an intervention in the normal course of events—if the abortion does not take place, the embryo will develop into a baby. Neither of these assumptions is applicable in the case of stem cell research. Consequently, one's views about the ethics of abortion do not determine one's views about the ethics of stem cell research. People opposed to abortion can be found on either side of the stem cell debate, as can people not opposed to abortion.

First of all, pregnancy, whether wanted or not, is not an issue. The embryos are grown in the laboratory and not in a woman's womb, so one party in the "abortion conflict" is not present at all. No one is challenging a woman's authority over her body and her pregnancy. Second, there is no issue of an intervention that would prevent these embryos from developing into babies. Left on their own, these early embryos would go nowhere. The only way we know to make these laboratory-grown embryos into babies is to implant them into a woman's womb, thereby making her pregnant. And even then, the chances that this will work are less than 25 percent. Because such a considerable amount of high-tech intervention is required, it is misleading to regard a laboratory-grown embryo as having the potential to become a baby. The necessity of high-tech intervention marks the difference between potential and merely possible. Otherwise we could say that each skin cell, because of cloning technology (which I discuss briefly below), has the potential and not merely the possibility of becoming a baby, which is absurd.

I do not wish to deny that there is some overlap between the ethical issues raised by stem cell research and the issues raised by abortion. My point is that the ethics of stem cell research should not be approached through the prism of the abortion debate. We need to identify and discuss the ethics of stem cell research in its own terms, in order to better understand what is at stake and to appreciate the odd alignment of advocates on each side of the debate. With that in mind, let us turn to the three ethical questions raised by this research.

The Moral Status of the Embryo

For most people, a baby has the moral status of a person—destroying a baby is a morally significant act of killing. For most people, sperm and eggs have the moral status of a piece of tissue—destroying tissue is not morally significant in itself. It is no more than cutting a fingernail or a strand of hair. The embryo is one stage in a developmental continuum, beginning with a sperm and egg and continuing through to a newborn baby. It is difficult, however, to think of moral status as a continuum. Hence, we seem to be faced with the question: Is destroying an embryo morally equivalent to killing a baby or to cutting a piece of tissue?

This question certainly seems similar to a question often raised in the abortion debate. And if one held that the embryo had the moral status of a person, that would seem to settle the matter so far as the ethics of stem cell research goes. Even if the research results in benefits that save many lives, it would be no more justified than the harvesting of vital organs from a healthy person in order to save the lives of several other people.

Interestingly, the issue of the moral status of the embryo has not been as crucial as it might be expected to be. The reason for this appears to be connected with the widespread acceptance of *in vitro* fertilization (IVF) as a treatment for certain kinds of infertility. This procedure consists of extracting an egg from the woman and sperm from the man and fertilizing the egg in a laboratory dish. After some days growing in the dish, the embryo is transferred to the woman's uterus with the hope that implantation will occur. Since the overall success rate of IVF (i.e., a live birth) is less than 25 percent, the standard practice in IVF is to medicate the woman so that she "superovulates," producing several eggs in a cycle. All of these are extracted and fertilized, and then inserted in the uterus, two or three at a time, until a successful implantation and pregnancy occurs. If success occurs before all the embryos are used, the remainder—the so-called "excess embryos"—are typically discarded. Despite papal condemnation of IVF, there has not been that much outcry or controversy over the procedure. Thousands of babies each year have been born as a result of IVF.

The current tolerance of IVF would not make sense if the moral status of the embryo were crucial. If thousands of live births from IVF occur each year, then several thousand excess embryos are discarded each year. Yet IVF clinics do not attract anything near the level of protests or controversy that abortion centers do. If one believes, as many of the strongest opponents to abortion do, that "life begins at conception"—i.e., that from the very moment of conception we are dealing with something that has the moral status of a person—then the annual destruction of thousands of excess embryos should be at least as offensive as the destruction of presumably far fewer embryos from stem cell research. Perhaps this difference reflects an inconsistency and the antiabortion movement should include IVF centers in their protests. Nevertheless, the fact that they have not suggests that the question of the moral status of the embryo may not be as decisive for the ethics of stem cell research as it might seem at first. Even if critics of stem cell research are inconsistent in their regard for embryos, we need an explanation for their inconsistency: Why should they consider the destruction of embryos in stem cell research as outrageous but the destruction of excess embryos in IVF as merely unfortunate?

Intentions

At least one commentator has argued that the difference between the destruction of embryos in IVF treatment and in stem cell research is a difference in the governing intention. The IVF clinician is creating embryos with the intention of enabling the woman to give birth to a baby. Even though excess embryos are usually destroyed, they were not created with that intention. The stem cell researcher, in contrast, is creating embryos with the intention of destroying them. This intention betrays an instrumental attitude towards embryos, marking a moral difference between the IVF clinician and the stem cell researcher. But perhaps we can avoid the presence of this presumably offensive intention in stem cell research: we simply restrict stem cell researchers to excess embryos. As others have remarked, there is no sense in letting these excess embryos go to waste.

Unfortunately, the supply of excess embryos from IVF clinics would not be adequate for stem cell research. Recall that one of the hopes for this research is the ability to replace dead or damaged cells associated with various diseases. Even if we are able to use stem cells to make cardiac tissue, we would still have the problem of compatibility—transplantable cardiac tissue grown from a stranger's stem cell will likely be rejected by the body. In order to ensure compatibility, the stem cells that produce the transplantable tissue should come from the very body that will receive the tissue. And in order to get pluripotent stem cells from a child or an adult, we would have to use somatic cell nuclear transfer—i.e., cloning. This technique rests on the surprising discovery of some years ago that transplanting the nucleus of a differentiated cell from an adult mammal—e.g., a skin cell—into an unfertilized egg cell, along with some further manipulations, can result in a totipotent stem cell. Cloning could therefore provide an adequate supply of compatible pluripotent stem cells.

Introducing cloning into the discussion of the ethics of stem cell research complicates matters, since cloning raises ethical concerns of its own. Creating a totipotent stem cell through cloning leads us down one of two roads: 1) We can develop that cell into an embryo with the intention of destroying it and harvesting its pluripotent stem cells; or 2) we can develop that cell into an embryo, fetus, etc., all the way to a live birth. The first procedure has been called "therapeutic cloning," the second "reproductive cloning." Most people have objected to reproductive cloning on the grounds that, with current technology, it is just too unsafe. Even if a live birth results, the risk of deformities or other genetic problems developing in the child is too high. For some, the worry about therapeutic cloning derives from the worry about reproductive cloning—permitting therapeutic cloning would increase the likelihood of someone engaging in reproductive cloning. But this isn't an ethical objection to the practice of therapeutic cloning itself; rather, it is a worry about the enforcement of restrictions on reproductive cloning if therapeutic cloning is permitted. Any practice or technology can be misused, but that is not sufficient reason to condemn it ethically. The stronger ethical concern about therapeutic cloning is the same as that associated with stem cell research: not only are embryos being destroyed, they are being created with the intention of being destroyed.

So we are back to intentions. Why is creating embryos with the intention of destroying them objectionable? The immediate answer is that this instrumental attitude regarding embryos displays a disrespect for human life, that condoning this intention puts us on a slippery slope that will lead us to treating human life as a mere commodity, to be sacrificed if the benefits are sufficient. The stem cell research team appears to have this troubling intention, whereas the IVF team does not.

But is there such an enormous difference between these two cases? Even though the IVF clinicians might create ten embryos for implantation, their hope is that they will succeed with the first attempt or two. They and the couple providing the gametes do not hope that they will have to use every embryo in order to achieve success. In other words, in IVF, embryos are created with the hope that most will not be needed. Thus, on the one hand we have the stem cell researcher who creates embryos with the intention of destroying them; on the other hand, we have the IVF clinician who creates embryos with the hope that most of the embryos will be unneeded—and so discarded. It is difficult to see a stark moral difference between the intention and the hope. If our condoning the hopes of IVF practice does not have the terrible consequence of fostering an instrumental attitude towards human life, then perhaps condoning the intentions of the stem cell researcher might not be so ominous.

Complicity

The politics of stem cell research has raised a third set of ethical issues. At one time the National Institutes of Health (NIH) tried to respond to the stem cell controversy by issuing a directive under which researchers were allowed to work with pluripotent stem cells so long as they themselves did not extract the stem cells and so destroy embryos. On a different occasion, President Bush issued an executive order permitting work on a few stem cell lines (i.e., a few stem cells and their descendants) that were developed before August 2001, but prohibiting any research on stem cells extracted after that date. Both cases—implicitly in the NIH directive, explicitly in the president's text—acknowledged that destroying embryos for research purposes is at least morally troubling. Each directive proposed a way of isolating and quarantining the problem so as to allow researchers to pursue the promise of stem cell research. If someone else could reliably supply the stem cells, in the case of the NIH, or if the sixty or so cell lines identified in the president's directive were adequate for research, then, it was hoped, we could have our moral scruples and swallow them too.

These directives raise the issue of complicity. Let us assume that destroying embryos for stem cell research is morally wrong. Does this thereby condemn any research or benefit derived from such destructions? In order to answer this, it is helpful to distinguish three levels of complicity. The first level is not actually engaging in the wrongful act but nevertheless providing incentives for others to do it. The second level is not encouraging others to act wrongly but nevertheless knowingly taking advantage of their actions when they do. The third level is, passively and perhaps unknowingly, enjoying the benefits of the wrongful act.

If destroying embryos for stem cell research is wrong, then clearly the people who actually engage in it deserve moral condemnation. Judging people who are complicit at various levels can be more controversial. The NIH directive seems to be an invitation to first-level complicity. NIH researchers would purchase stem cells from outside suppliers who would engage in the actual destruction of embryos. Some people, presumably certain NIH officials, might believe that this policy leaves the NIH researcher with clean hands. Nevertheless, many would not think much better of the person who was complicit at this first level, on the theory that the person hiring the assassin is no more innocent than the assassin himself. The president's directive seems to be an invitation to second-level complicity. Since the cutoff date preceded the date of the directive, it could not provide an incentive or encouragement for any further destruction of embryos. And yet, although these past destructions were presumably wrong, the presidential directive invites researchers knowingly to take advantage of these wrongful acts. One should not regard people complicit at this second level as completely guilty or as completely innocent, though this assessment will vary with the circumstances and gravity of the wrongful act. Someone who pockets some cash inadvertently dropped by a fleeing bank robber is one thing; a scientist who uses data murderously obtained by Nazi doctors is quite likely another. In any case, inviting people to be complicit at the second level is itself troubling: it would seem to hypocritically subvert the moral condemnation of the original wrongful act. And it would not allow for an adequate supply of stem cells, as discussed above. Although presented as a compromise, it produces little benefit and some harm.

It is sometimes easy to forget that the debate over stem cell research in the United States is a somewhat parochial matter. Prohibiting stem cell research in the US would not stop such research altogether; it is not clear that it would even seriously hamper or slow it down. For while the matter is debated in Congress, Britain has already decided to allow the use of embryos in research as long as they are less than fourteen days old. And Britain is no second fiddle when it comes to biological research: two of the most important discoveries in biology in the past 150 years—the theory of evolution and the structure of DNA—occurred in England. The discovery that mammalian cloning was possible occurred in Scotland. Hence, given the present circumstances, we should not be at all surprised if many of the important discoveries regarding stem cells also occur in Britain. If this happens and if clinical benefits are then forthcoming, we may well wonder what sort of debate will arise in the US over their importation. Would the people currently opposed to stem cell research argue against or excuse people who wish to enjoy the clinical benefits—the third level of complicity? It's clear what the response of the public would be.

Sources

George W. Bush's August 9, 2001 statement can be found at: http://www.whitehouse.gov/news/releases/2001/08/20010809-2.html; the commentator who argued that the difference between the destruction of embryos in IVF treatment and in stem cell research is a difference in the governing of intention, see: Charles Krauthammer, "Crossing Lines: A Secular Argument against Research Cloning," *New Republic*, vol. 29 (April 2002) and also see http://www.tnr.-com/doc.mhtml?i=20020429&s=krauthammer042902&c=1; the National Institutes of Health directive that allows researchers to work with pluripotent stem cells can be found at: http://www.nih.gov/news/stemcell/stemcellguidelines.htm (the relevant section is II.A.2).

What Makes Genetic Discrimination Exceptional?

Deborah Hellman

Introduction

Recent advances in understanding the genetic basis of disease have inspired hope but also have led to fear. Scientists, physicians, and genetic counselors, along with their patients and potential patients, worry that those whose genetics place them at risk for a serious disease will face discrimination. Denial of health insurance is one concern, but people also fear discrimination in such other aspects of life as employment and child custody decisions. In response to these concerns, many state legislatures have passed laws forbidding genetic discrimination. Most legislation of this sort addresses discrimination in health insurance, but some legislation is directed to employment, or to life or disability insurance.

These laws have been met with both praise and criticism. Defenders of the laws see them as important and necessary—even if they do not go far enough. Critics view such legislation as unjustified and unwarranted. After addressing an important preliminary issue: the problem of defining "genetic discrimination" in a way that adequately differentiates it from health status discrimination more generally, I turn to the central question of genetic exceptionalism. That is, I ask: Is genetic discrimination different from discrimination on the basis of health such that it warrants special protective legislation? To do this, I examine—and reject—most familiar arguments in support of genetic exceptionalism. However, one of these arguments merits further consideration. If genetic discrimination discourages individual participation in research and treatment, then opportunities to gain further knowledge beneficial to the wider public would be lost. Legislative protection might be justifiable in order to remove barriers that genetic discrimination places on medical advances, which are essential in the promotion of the general welfare.

A second argument that warrants a closer look implicitly underlies many of the arguments in the literature. Genetic discrimination might be meaningfully different (and worse) than health status discrimination because of what it expresses. Many commentators refer to the history of eugenics in this country and elsewhere without clearly articulating why that history matters. I try to fill that gap by developing the argument that, because the social meaning of treating people differently on the basis of their genetic makeup is different from the social meaning of discrimination on the basis of health or illness, special legislation to prohibit genetic discrimination is warranted.

Defining "Genetic Discrimination"

Some who oppose prohibitions on genetic discrimination believe that it is theoretically or practically impossible to distinguish genetic discrimination from discrimination on the basis of health more generally. These critics argue that if genetic discrimination is legally defined as discrimination based on information derived from a test of a person's genetic material—an examination of DNA, for example—then laws will fail to capture many instances of discrimination on the basis of genetic predisposition to disease. For example, a family medical history—the common starting point of any medical record or doctor's visit—contains a wealth of information about a person's genetic makeup. Critics thus contend that state laws that define "genetic information" as information resulting from a test of DNA are overly narrow. In response, newer laws define genetic discrimination more broadly. However, these laws inadvertently seem to prohibit almost all forms of discrimination on the basis of health (except perhaps illness or disability caused by accident). As many diseases are at least partly influenced by our genes, common tests ordinarily not considered "genetic tests"—a blood pressure reading, for example—arguably could fall within the purview of these new laws.

Medical ethicist Henry Greely offers a helpful resolution to this definitional problem. In his measured support for limited federal legislation prohibiting genetic discrimination in health insurance and employment, Greely suggests that health insurers should be prohibited from denying coverage or charging higher premiums based on information about genotype (unexpressed genetic traits)—no matter the source of that information. As he explains, "[g]enetic information should thus be defined broadly to encompass any . . . information that provides probabilistic information about a person's genotype . . . from genetic tests, other medical tests, family history, diagnoses of traits or conditions, or the taking of (or even making inquires about) a genetic test." For Greely, once a genetic predisposition is manifest (phenotype)—as illness or at least as a medically relevant symptom—it becomes a matter of one's health status, and discrimination would not be prohibited.

This distinction between discrimination based on unexpressed genetic traits and discrimination based on manifested illness takes us to the point where we can evaluate arguments in support of genetic exceptionalism. That is, we must consider whether genetic discrimination is different from other forms of health status discrimination such that it warrants special legislative attention.

Genetic Exceptionalism: The Familiar Arguments

Most Americans with health insurance are covered through group rated plans—that is, no individual risk assessment is made at the time of their request for coverage. However, individual risk assessment is common for those who buy *individual* health insurance policies as well as in the life, disability, and long-term-care insurance markets. At present, a person who is sick may be charged higher insurance rates or denied coverage. Why then forbid similar treatment of someone with a genetic predisposition to the very same illness?

Some might say that drawing the distinction between sick and healthy individuals is "fair," while distinguishing between two healthy persons (only one of whom carries a genetic mutation that predisposes him to disease), is "unfair." The sick person is already sick, so the thinking goes, and surely will need to make an insurance claim (perhaps many claims); the healthy person with a genetic *predisposition* to an illness might never need to make a claim. This intuition is misguided however. The sick person will not *necessarily* need health care services—she might be hit by a bus and killed on the first day of the policy period. Moreover, some with a genetic predisposition to an illness *will* develop the predicted illness, so long as they do not die of something else first. Insurers make predictions about the likelihood that claims will be made; the difference between the person who is already sick and the person who *might* become sick is a difference in degree and not in kind.

Although these considerations focus on insurance, the question whether genetic discrimination is different, morally speaking, from discrimination on the basis of health or illness also can be raised in other contexts. Employment discrimination has attracted attention; some states have passed legislation forbidding genetic discrimination in employment. Here too, then, one must ask whether genetic discrimination in employment ought to be specifically forbidden.

Genetic discrimination is irrational. Some argue that legislative prohibitions on genetic discrimination are necessary because genetic discrimination is irrational. In the case of insurance, irrational discrimination would occur if the insurer charges higher rates to a group that is not in fact more likely than average to make a claim. Rational discrimination, by contrast, makes distinctions among groups in a way that reflects the *real risk* of loss posed by each group. More precisely, discrimination is irrational if the prices charged by the insurer do not reflect the actual risk of loss posed by each group *as well as the cost to the insurer of distinguishing between the groups.* Understood in this way, it is unclear what is morally important about the irrational character of discrimination. Irrational discrimination is simply synonymous with bad business.

Further, while it is true that many who carry a gene that predisposes them to a particular illness will not in fact become sick, it is also true that such persons are *more likely than average* to develop that illness. Thus, discrimination on this basis may well be rational. Moreover, the probabilistic nature of genetic information is no different from other information about a person's future health used by insurers to set rates. Not all smokers develop lung cancer, but because they are more likely than average to develop lung cancer, it is rational to charge smokers higher health insurance rates.

Some argue that genetic discrimination is more complex and less predictive of future health than smoking is of cancer. If insurers do not understand this complexity, they might discriminate in ways that are irrational. Perhaps. But if so, is this a problem that requires a legislative solution? Insurance statutes of all states already require that rates be grounded in actuarial data; state law generally requires that insurance rates be rational. In employment law, by contrast, there exists no general requirement of rational behavior. If an employer wants to discriminate irrationally, hiring only brown-eyed applicants, for example, such irrational discrimination by itself is not prohibited. Irrational discrimination of a *special sort* is prohibited, but not *because* it is irrational. Race, sex, and disability discrimination, for example, are largely prohibited both when they are rational and when they are irrational.

At present, there is little evidence of genetic discrimination. However, were it to become a problem, three reasons would argue against prohibition. First, such a law would be overinclusive if its aim is to ban only irrational genetic discrimination, since some genetic discrimination is rational. Second, current law already bans irrational discrimination in insurance, though these laws are less potent than would be a law *specifically* banning genetic discrimination in insurance. Third and most importantly, it is not clear why being subject to irrational discrimination is a significant moral harm that requires remediation. Rational discrimination is simply the making of distinctions that are economically sensible, according to the insurer, employer, or other actor who draws such distinctions. A bad business judgment, without more, does not constitute a moral wrong to the person disadvantaged by that judgment.

Genes are beyond individual control. A common argument for singling out some attributes for protection from discrimination is that they are "immutable," or beyond individual control. The moral intuition underlying this argument is that a person ought to be granted or denied benefits on the basis of what she *does,* and not who she *is.*

Immutability fails as a reason to prohibit genetic discrimination for two reasons. First, most goods are distributed according to principles that often have little or nothing to do with what one does. The basketball player who earns millions of dollars for his performance earns that money only partly in recognition of his effort; his height, surely beyond his control, and his natural talent also play a role in making him a skilled player. Second, while the notion of personal responsibility has moral appeal, it is far more complex conceptually than is immediately apparent. We might say that a smoker ought to pay high insurance rates because, in choosing to smoke, she is partly responsible for her greater risk of illness. But even that example is problematic because of the additive quality of nicotine and the fact that individual differences in our bodies affect the degree to which smoking endangers health. Our bodies make demands on us unevenly.

A small number of people are especially burdened by genetic disease. A related argument is based on the recognition that genetic predisposition to disease is distributed quite unequally. While most of us may turn out to have roughly equivalent risk profiles—a higher than average risk of X, but a lower than average risk of Y—smaller groups of people are extremely lucky or extremely unlucky in their genetic make-up. Social policies that permit discrimination on the basis of genetics would seem to cruelly compound their misfortune. However, although this argument initially seems persuasive, it does not explain why those whose genetic makeup render them especially likely to *become* sick deserve special solicitation when those whose poor health is already manifest do not.

Genetic traits overlap with racial or ethnic groups. One reason that the genetically unlucky may deserve special protection is because genetic predispositions affect identifiable racial or ethnic groups. For example, sickle-cell anemia generally affects Africans and African Americans; two mutations associated with breast cancer are more common among Ashkenazi Jews than others. Where the group affected is already stigmatized in our society, there is a risk of further entrenching negative attitudes and of overreacting to the significance of the association.

Although these associations are important, they do not justify a ban on genetic discrimination. Although the connection between some genetic diseases and stigmatized racial or ethnic groups such as African Americans or Jews seem significant today, this significance will probably wane. The nature of genetic research makes it easier to identify genetic mutations among relatively homogeneous and relatively small ethnic groups. But as scientists are able to identify the function of a greater number of genes, the relevance of small population studies will dissipate. Moreover, there are many disease-causing mutations that are not more prevalent among stigmatized populations—Huntington's disease and early onset Alzheimer's disease are two particularly devastating examples. To ban genetic discrimination because of the risk of the further stigmatization of racial and ethnic groups would be to enact a law that is quite dramatically both over- and underinclusive.

Stigma. Some argue that genetic predisposition to disease is *itself* stigmatizing. Similar to the concern that racial prejudice creates a color hierarchy in our society, some worry about a genetic hierarchy. For example, those with genetic predispositions to diseases may become less desirable as customers for insurance, as employees, or as marriage partners.

While superficially appealing, this argument is underdeveloped. First, one must claim not just that possession of a genetic predisposition to disease is stigmatizing, but that it is *more* stigmatizing than having an already manifest illness. More importantly, it is not clear what is meant by "stigma." Perhaps "stigma" refers to the *effect* on the persons with the genetic condition. If so, genetic discrimination is wrong because it causes harm. But questions remain: What harm in particular is at issue, and why does *this* harm render discrimination wrong? If someone is denied a job or insurance coverage because of genetic traits she carries, she *is* certainly harmed, but it is not a *stigmatic* harm. Perhaps stigmatic harm refers to the psychological effect on the person denied the good because of her genetic traits. This way of understanding stigma is familiar from arguments about what makes race discrimination wrong. But if harm to the person subject to discrimination were a necessary component of wrongful discrimination, then racial segregation of facilities for those in a permanent vegetative state would not be wrongful, because such persons are incapable of suffering psychological or emotional hurt. As this conclusion seems untenable, the claim that wrongful discrimination requires that those affected feel stigmatized fails.

Finally, "stigma" might refer to what the policy of genetic discrimination *expresses*. Perhaps genetic discrimination is wrong because of the meaning expressed in distinguishing people on this basis. (This expressivist argument is implicit in much of the critique of genetic discrimination. It has not, however, been carefully articulated and evaluated, a task I take up in the discussion, "Expressivism and Genetic Discrimination," below.)

The notion of stigma also carries with it the idea of a class or caste-like distinction between groups of people. Perhaps what is wrong with genetic discrimination—which distinguishes it from discrimination on the basis of health—is that it threatens to create a *genetic underclass*. This fear motivates an argument for legislation prohibiting genetic discrimination that has been most forcefully articulated by the philosopher Susan Wolf.

"Geneticism." Susan Wolf argues that the tendency to focus on genetics and to subordinate people on that basis is best termed "geneticism"—a term that she uses to call attention to a deeply ingrained mindset and set of structural practices. Her view is grounded in critiques of the antidiscrimination approach as applied to problems of racism and sexism. Similar to those cases, she believes that "clinging to 'genetic discrimination' . . . creat[es] a false genetic 'norm,' frustrating structural reform, obscuring the deep psychological roots of genetic stereotyping and prejudice, and isolating genetic from other harms." Wolf's conception of the issue as something deep and entrenched leads her to argue that the eradication of geneticism requires a systemic approach.

Thus, while Wolf explicitly does not support laws prohibiting genetic discrimination, her approach does support the idea of genetic exceptionalism. But Wolf fails to address in what way the particular inequalities identified by *geneticism* are morally problematic. To complete her argument she must also claim that *this inequality* clashes with the commitment to treat each person with equal concern (or some similar formulation of a general principle of equality). As economist and philosopher Amartya Sen insightfully emphasizes, since equality among people across all dimensions of life is impossible, moral theories differentiate themselves by articulating what sort of equality is morally significant. To merely note a particular inequality—people with trait X have less of Y—is not enough. One must also explain *why* this sort of inequality is one that is morally problematic.

Promoting Health

The scientific advances in understanding the genetic basis for disease have enormous potential to improve health. Understanding individual predispositions to disease might allow medical advice to be individually tailored, both for prevention and for treatment. Most exciting of all, a greater understanding of genetics could allow doctors to treat people with strong predispositions to serious illness prophylactically so that the illness itself never develops. Genetic discrimination is a problem, then, because it may get in the way of each of these beneficial developments. If people fear genetic discrimination, they may be reluctant to be tested for genetic conditions. If so, achievement of the health benefits described above may be thwarted in two ways. First, if people avoid testing, they may fail to partake in the therapeutic benefits that are currently available. Second, and perhaps more importantly, people who fear discrimination may decline to participate in research involving testing—research that could lead to discoveries that ultimately make presymptomatic treatment of genetic conditions possible. This argument is important but rests on several empirical assumptions that require further consideration.

In order for legislation forbidding genetic discrimination to be successful in removing this barrier to testing, the legislation must forbid the sorts of genetic discrimination that people actually fear. If people also fear discrimination in life, disability and long-term-care insurance, for example, legislation that is limited to the health care and employment contexts is unlikely to promote health. Consider the case of someone contemplating enrollment in a research study dealing with the genetic predisposition to early onset Alzheimer's disease. For this person, the availability of long-term-care and disability insurance are critical issues. People's fears of genetic discrimination might extend further still. Perhaps their central worry is that they will be unable to adopt children, or that in a custody battle they will be denied custody of the children because of their genetic condition, or that mates or family members will abandon them. Research to date suggests that people's fears are not assuaged by protective legislation, but more research is needed to fully understand whether legislation can affect decisions about whether to undergo genetic testing.

Expressivism and Genetic Discrimination

Much of the commentary about genetic discrimination refers to the history of eugenics both in this country and in Europe. It seems obvious that this history is relevant to the question of whether genetic discrimination is wrong as well as to whether genetic discrimination is different from discrimination on the basis of health such that it warrants special legislation. Exactly why this history matters, however, is not clear from the arguments currently found in the literature. The expressivist argument fills that gap; the history of eugenics is relevant because it changes the social meaning of current practices. Genetic discrimination expresses something different because of our experience with illegitimate uses of genetics.

The expressivist account is an especially close cousin of the argument that genetic discrimination is wrong because it is stigmatizing. Although "stigma" is an elusive term, it is fair to say that it generally calls attention to the *effect* of a law or policy. An expressivist approach, in contrast, focuses on what is *expressed* by a law or policy—regardless whether this expression actually harms an identifiable group in a particular way.

Expressivism contends that what an action expresses—its meaning—is relevant in determining its moral permissibility. In contrast, moral permissibility typically depends on the *intent* of the actor, the *effect* of the action, or both. For example, in criminal law both intent and effect are relevant in judging the moral acceptability of an action. The person who kills another by accident will not likely be guilty of murder, which requires that the actor have a particular intent. In addition, if someone intends to kill another but fails, she may be guilty only of attempted murder and thereby be punished more lightly because her action resulted in no significant harm. To claim that the expressive character of an action is relevant is to call attention to a third dimension of action: my spitting on a homeless person, for instance, is wrong because it *expresses* disrespect. In this example, the *effect* of the action (and the experience of being spit on is certainly unpleasant) clearly is far less important than what it *expresses.* Further, if I intend disrespect by spitting, an account of intent alone cannot explain why—or the depth to which—spitting is disrespectful. And even if I do not intend disrespect by spitting, I cannot spit blamelessly: the *meaning* of spitting on someone in our culture is as an act of disrespect. The expressive dimension of action allows for assessment of moral permissibility that is not reducible simply to consideration of intentions or consequences.

The history of eugenics chronicles the misuse of genetic information, and colors the way that we—our society and people—understand the practice of genetic discrimination. Although the beginning of the systematic, scientific understanding of genetics and its effect on health stretches back about 150 years, it was not until the twentieth century that these scientific developments began to have significant social consequences. Some scientists and social reformers viewed the promise of genetics as the power to affect reproductive choices—eugenics. Eugenicists wanted to encourage those perceived as having "better" genetic endowments to reproduce more, and toward that end, for example, in the 1920s state fairs around the country sponsored "Fitter Families" contests, with prizes for "Grade A individuals" in the "human stock" category. This interest in encouraging the "better" to reproduce was coupled with a fear that the "defective" were over-producing. In the US, the eugenics movement had racial overtones as well, with "white Protestants of Northern European Stock" superior to "blacks and Jewish and Catholic immigrants." It was not until the Second World War, with revelations of Nazi atrocities, that the use of genetic science for eugenic purposes was discredited.

Read with this history as a backdrop, the meaning of genetic discrimination may be that those with genetic flaws (or more flaws than average) are less worthy or less important— *even if* those genetic aspects never see expression—that is, genetic discrimination is directed at genotype (not phenotype). If this understanding of the meaning of genetic discrimination as acceptance of moral inferiority is correct, then genetic discrimination is different from discrimination on the basis of health and, furthermore, genetic discrimination is morally impermissible. This argument's strength depends on the claim that genetic discrimination does express a meaning that denigrates the equal moral worth of those with genetic predisposition to disease. Interpreting the social meaning of a policy or practice is a difficult interpretive task about which reasonable people will often differ. Expressivism's contribution is the suggestion that we ought to turn our attention to this issue.

Conclusion

Recall that, to distinguish genetic discrimination from discrimination on the basis of health, I adopted Henry Greely's proposed solution, which suggests that genetic discrimination should be defined as discrimination on the basis of information about genotype regardless the source of that information. However, once a person manifests illness (regardless whether it is of genetic, nongenetic, or mixed origin), then discrimination on this basis would not be considered *genetic discrimination*. I suggest that genetic discrimination, so defined, might be considered meaningfully different from health status discrimination if it *expresses* denigration of people with genetic disease. When discrimination is based on asymptomatic genetic predispositions or conditions, it is likely that such discrimination will be seen as *genetic*, thereby calling up the history of eugenics. In addition, genetic discrimination might warrant special legislation in order to fulfill the promises of genetic science.

Many questions remain, however. First, we must reach a better understanding of how legislation affects testing decisions. Second, we must continue the discussion, in public and private, about what, if anything, genetic discrimination expresses and in particular whether it denigrates the equal worth of people with genetic disease.

A fuller articulation of this view can be found in "What Makes Genetic Discrimination Exceptional? *American Journal of Law & Medicine*, vol. 29, no. 1 (Spring 2003), pp. 77-116; this article appears with the permission of the American Society of Law, Medicine & Ethics and Boston University.

Ryan Lemmerbrock deserves acknowledgement and thanks for his excellent research assistanance.

Sources

For the state legislatures that have passed laws forbidding genetic discrimination in such areas as health insurance, employment, or life or disability insurance, see: National Conference of State Legislatures (NCSL), *State Genetic Discrimination in Health Insurance Laws,* (June 2001), *available at* http://www. ncsl.org/programs/health/genetics/ndishlth.htm (last modified April 3, 2002) and also NCSL, *State Genetics Employment Laws, available at* http://www.ncsl.org/programs/health/ genetics/ndiscrim.htm (last modified April 29, 2002); for an example of a defender of laws against discrimination, but who also does not believe those laws go far enough, see Trudo Lemmens, "Selective Justice, Genetic Discrimination, and Insurance: Should We Single Out Genes In Our Laws?" *McGill Law Journal,* vol. 45 (2000); critics of legislative prohibitions against genetic discrimination include Colin S. Diver and Jane Maslow Cohen, "Genophobia: What Is Wrong with Genetic Discrimination?" *University of Pennsylvania Law Review,* vol. 149 (2001); for an example of a critic of prohibitions on genetic discrimination, because it is theoretically or practically impossible to distinguish genetic discrimination from discrimination on the basis of health more generally, see Thomas Murray, "Genetic Exceptionalism and Future Diaries: Is Genetic Information Different from other Medical Information?" in *Genetic Secrets: Protecting Privacy and Confidentiality in the Genetic Era* (Yale University Press, 1997); Henry T. Greely has argued that state laws defining "genetic information" as information resulting from a test of DNA have applied an overly narrow definition, see: "Genotype Discrimination: The Complex Case For Some Legislative Protection," *University of Pennsylvania Law Review,* vol. 149 (2001). One instance of an overly broad application of "genetic information" is the Maryland statute, which provides: "[a]n insurer, nonprofit health service plan, or health maintenance organization may not: use a genetic test, or the results of a genetic test, genetic information, or a request for genetic services to reject, deny, limit, cancel, affect a health insurance policy or contract." Md. Ins. Code Ann. § 27-909(c)(1)(2002). The phrase "genetic exceptionalism" was first coined by Thomas Murray; for an example of arguments that legislative prohibitions of genetic discrimination are necessary because genetic discrimination is irrational, see: Larry Gostin, "Genetic Discrimination: The Use of Genetically Based Diagnostic and Prognostic Tests by Employers and Insurers," *American Journal of Law & Medicine,* vol. 17 (1991); the discussion of rational discrimination as reflecting real risk of loss posed by an insured group occurs in Kenneth S. Abraham, *Distributing Risk:*

Insurance, Legal Theory, and Public Policy (Yale University Press, 1986); for an argument that some attributes deserve protection because of their immutability, see: Erwin Chemerinsky, *Constitutional Law* (Aspen Law & Business, 2001); for an argument is that a person ought to be granted or denied benefits on the basis of what she *does,* and not who she *is,* see: Judith Lichtenberg and David Luban, "The Merits of Merit," *Report from the Institute for Philosophy and Public Policy,* vol. 17 (1997); on negative attitudes as a result of genetic traits shared within a racial or ethnic group, see: Janet L. Dolgin, "Personhood, Discrimination and the New Genetics," *Brooklyn Law Review,* vol. 66 (2000–2001); Susan M. Wolf, "Beyond 'Genetic Discrimination': Toward the Broader Harm of Geneticism," *Journal of Law, Medicine and Ethics,* vol. 23 (1995); Amartya Sen, *Inequality Reexamined* (Clarendon, 1992); Mark A. Hall and Stephen S. Rich, "Genetic Privacy Laws and Patients' Fear of Discrimination by Health Insurers: The View from Genetic Counselors," *Journal of Law, Medicine and Ethics,* vol. 28 (2000); for a discussion of the relevance of the expressive dimension of action, see *Maryland Law Review,* vol. 60 (1999) (symposium issue); Daniel J. Kevles, *In the Name of Eugenics: Genetics and the Uses of Human Heredity* (Knopf, 1985).

My Fair Baby: What's Wrong with Parents Genetically Enhancing Their Children?

David T. Wasserman

In the shadow of state-sponsored eugenics, much of the debate over the still-distant prospect of human genetic enhancement concerns the role of the government. Some fear that the state will use genetic technology to impose its own conception of the good life or to suppress or neglect minority conceptions; others fear that the state will not adequately regulate the genetic marketplace, allowing parental choices of desired traits to have cumulatively adverse effects, e.g., creating substantial sex imbalances, reinforcing narrow, oppressive ideals of beauty or excellence, and compounding social inequality. Without slighting these concerns, I will focus in this paper on the ethics of individual parental choices. (I will also ignore the important but distinct issue of parental complicity in the adverse social effects of human genetic enhancement.) Although I will argue that many objections to parental enhancement are confused or incomplete, I will suggest that enhancement by parents may raise different moral concerns than enhancement by the state and other third parties. Specifically, I will consider whether there are special moral problems in attempting to shape in advance the talents, preferences, and values of an individual with whom one expects to enter into a special kind of intimate relationship.

Presuming Prenatal Consent

I will begin with a set of objections that apply directly to parents, although they are not limited to them. The general principle underlying these objections is, roughly, that if we do something to a person's body (or to the body that will become the person's) that may significantly affect or potentially harm her, we must either obtain her consent, or, if that is not feasible, be able to reasonably presume her consent.

The requirement of informed or presumed consent, a mainstay of current medical practice, is sometimes invoked to reject genetic alterations to unborn children or future generations. Since they cannot consent, and we know nothing about their willingness to incur risk, we must not impose any change that carries some risk. But this would rule out virtually any genetic intervention, even ones that would impose a very slight risk of avoiding minor harms for a very substantial reduction in the risk of avoiding major ones. It would rule out any life- or health-preserving prenatal or germ-line therapy that was not perfectly safe, which is to say any such therapy. One way of avoiding that result would be to adopt a less risk-averse standard. While it would not be feasible to establish precise criteria, because of the familiar difficulties in quantifying risk, we might develop rough guidelines that reflected societal or cultural norms about risk.

Critics of genetic enhancement argue, however, that while we can rely on such norms to justify therapy for unborn or unconceived children, we cannot rely on them to justify enhancement. The reason they give is much like the standard one offered for the priority of health care as a social good—that health and longevity are goods that people want whatever else they want, and that morbidity and mortality are correspondingly universal evils. Proponents of enhancement counter that strength and intelligence are no different in this respect than health and longevity. The former have the same general utility as the latter, and they likewise expand an individual's opportunity range, whatever his goals and values.

Some critics respond that we can presume consent to risks or harms imposed to avoid harms, but not to confer benefits, however general—a surgeon can break the arm of an unconscious accident victim to save his leg but not to raise his IQ. While we can presume a willingness to trade off lesser against greater harms, we cannot presume a willingness to trade off harms against benefits. The acceptability of the latter kind of trade-off depends too much on the individual's own preferences, values, and ends to justify a presumption.

This response, however, merely exposes the problem in applying the notion of presumed consent to unborn and unconceived children. That notion was developed out of, and derives its moral force from, a commitment to deferring, to the greatest extent possible, to the preferences, values, and ends of existing people unable to choose for themselves. The problem is not that we lack sufficient information about the embryo's preferences, values, and ends, as we might in the case of a solitary elderly patient now in a comatose state. The early embryo simply does not have such preferences, values, and ends, and it makes no sense to infer them from biographical evidence or social generalizations, as we might for an elderly comatose patient. If we shift from presumed to hypothetical consent, we avoid the false analogy of the unborn to the unconscious, but we land in a briar patch of controversy on the identity and knowledge of those asked to consent, and the binding effect of their consent—a controversy from which we are unlikely to emerge with clear answers. And if we shift to retroactive consent—the consent of the already enhanced—we face the obvious problem that such consent may result directly or indirectly from the interventions at issue, and thus lack moral authority.

Enhancement as Autonomy-Infringing

A close cousin to the objection based on the presumed consent of the future child is the objection based on its autonomy. It is a weaker objection, since it would not apply to such general enhancements as heightened strength or intelligence, but only to a range of enhancements that would more narrowly shape the child's projects and attachments: the enhancement of specific attributes, such as height or musicality, that differentially promote a narrow set of life plans, plans the parents favor. The question now becomes whether such enhancements can be said to infringe the autonomy of the future child endowed with them.

Admittedly, specific enhancements would not expand a child's opportunity range as much as general enhancements. But no modification we regarded as an enhancement would be likely to substantially narrow the child's opportunity range. For example, if an early embryo likely to be of average height were endowed with the height potential for NBA play, he would gain opportunities in basketball, lose opportunities in horse racing, but very likely enjoy a net increase in opportunity in a society that values height. (On the other hand, a child genetically endowed with short stature for a career as a jockey should arguably not be regarded as enhanced, given the myriad disadvantages of being short in our society.) Moreover, our tolerance of a wide variety of opportunity-constricting postnatal interventions, from a strict, insular religious upbringing to a rigorous childhood regimen of musical or athletic training, suggests that we believe that a child's autonomy is not compromised, except in extreme cases, by limited opportunities.

A variant on the claim of opportunity limitation is the claim of manipulation. It might be claimed that a child endowed with specific enhanced abilities will be likely to develop compatible motivations and interests, and that by promoting this outcome, his parents violate his autonomy by manipulating him. Political theorist Matthew Clayton quotes legal philosopher Joseph Raz to support this view: "Manipulation, unlike coercion, does not interfere with a person's options. Instead, it perverts the way a person reaches decisions, forms preferences, or adopts goals. It is an invasion of autonomy whose severity exceeds the importance of the distortion it causes" [citation omitted]. But endowing a person with unusual height appears to do no such thing. An adolescent genetically engineered to reach seven feet could make decisions, form preferences, and adopt goals in a normal manner, trying to match his interests and abilities. The onus is on the critic to explain why the fact that those abilities had been genetically enhanced would pervert his decision-making processes.

It would be difficult to find such perversion even if his parents somehow managed to endow him with psychological attributes predisposing him to athletics. The fact that he would owe not only his physical ability but his psychological predispositions to someone else's plans for him would still leave him free to reach decisions in the same way as an individual who came by his abilities and predispositions in more familiar ways. Again, the onus is on the critic to explain how the parents' role in shaping his preferences prenatally would pervert his decision-making process, especially in light of the fact that parents often, and unobjectionably, attempt to shape their children's preferences after birth.

Selectivity versus Control

The most plausible objection to genetic enhancement may be that to shape the nature of one's own child before it is born or even conceived is to begin an intimate relationship in an improper manner. This objection does not rest on the psychological claim that one who enters an intimate relationship for bad reasons is unlikely to fulfill its responsibilities. We can recognize the transformative power of some relationships—to deepen the commitments with which they begin, and to make better people out of their participants—and still insist that it is wrong to begin those relationships in certain ways. Some adult relationships, for instance, begin in lust or opportunism and mature into love or friendship, and we can approve the end result while still disapproving the conduct that made it possible.

It might seem, however, that we would not have the same grounds for objecting to the conduct of parental genetic engineers. In contrast to seducers and opportunists, those parents do not do anything wrong to the early embryo; they do not, for example, deceive or exploit it, because it is not at that stage a being that can be deceived or exploited. Yet the way in which they initiate the relationship may still convey a lack of respect for the dignity of the future child or the parent-child relationship. But while we have secular accounts of the wrongs in casual seduction and opportunistic friendship, it is not clear what, if anything, would be correspondingly wrong in parental efforts to shape a child's nature prenatally.

There are two possibilities. One is to assimilate the wrong of genetic enhancement to the wrong of prenatal or preconception screening. Perhaps the objection is to the kind of finickyness or snobbery that can manifest itself in the selection of children as well as in the selection of friends or spouses. And perhaps such finickyness or snobbery is even more offensive in selecting children than in selecting friends or spouses.

It might be argued that parents would display objectionable snobbery in selecting or creating a child with superior attributes even if, with more limited options, they would accept a child with inferior attributes. This claim can be illustrated by taking adoption as an intermediate case between choosing friends and making children. If adoption agencies allowed a prospective parent to select a child on the basis of its known attributes—which they generally don't (except, to some extent, as a dubious incentive in the case of hard-to-adopt children)—parents who would be willing to adopt a child of ordinary intelligence with no other choice might still display objectionable snobbery if they preferred one with extraordinary intelligence, and would chose one with that attribute if they could do so. Similarly, parents who would use preimplantation genetic screening to select an embryo likely to develop extraordinary intelligence might display objectionable snobbery even if, in the absence of screening, they would accept an embryo lacking that potential. Parents who enhance an early embryo genetically might also display such snobbery, but they need not—they might be committed to the specific embryo they conceived or implanted, and might refuse to substitute one that already had the desired genetic endowment. If parents who genetically enhance their future children act wrongly, it is not only, or primarily, in displaying finickyness or snobbery.

This suggests that there may be distinct concerns about creating desired traits in children and selecting for them in friends, or even in children. The more apt objection to genetic enhancement may concern excessive control rather than excessive selectivity. The challenge is to explain how control can be a vice when it is exercised in shaping a fetus, rather than in overbearing the will of a conscious, autonomous person. What is wrong with profoundly shaping a person with whom one intends or expects to have an intimate relationship? After all, parents and teachers do it all the time.

Fearful Asymmetry

The moral difference may lie in the fact that genetic enhancement, unlike rearing and education, would be unilateral rather than collaborative. But much child rearing is done when the child is not even self-conscious, let alone capable of meaningful collaboration. And no genetic enhancement would be likely to bear fruit unless at some stage, the parents were able to enlist their child's collaboration through rearing and education.

The philosopher Jürgen Habermas makes some suggestive remarks on this point. He claims that genetic "programming" would make the normal asymmetry of the parent-child relationship irreversible in a way that rearing and education do not. This difference would not be apparent if the child were inclined to adopt the plans his parents have made for him; it would only emerge if he were inclined to question or reject those plans:

> Of course, the adolescent may assimilate the "alien" intention which caring parents before his birth associated with the disposition to certain skills much in the same way as might be the case, for instance, for certain vocational traditions running in the family. For the adolescent confronted with the expectation of ambitious parents to

> make something out of, for instance, his mathematical or musical talents, it makes no fundamental difference whether this confrontation takes place in reflection on the dense tissue of domestic socialization or in dealing with a genetic program, provided he appropriates those expectations as aspirations of his own. . . .

In cases where the parents and child have "dissonant intentions," however, the moral difference between genetic enhancement and socialization becomes clear:

> Due to the interactive structure of the formation processes . . . , expectations underlying the parents' efforts at character building are "essentially contestable." . . . [T]he adolescents in principle still have the opportunity to respond to and retroactively break away from [parental rearing]. . . . But in the case of a genetic fixation carried out according to the parents' preferences, there is no such opportunity. . . . Being *at odds* with the genetically fixed intention of a third person is hopeless. The genetic program is a mute and, in a sense, unanswerable fact. . . .
>
> Eugenic programming establishes a permanent dependence between persons who know that one of them is principally barred from changing *social* places with the other. But this kind of social dependence, which is irreversible because it was established by ascription, is alien to the reciprocal and symmetrical relations of mutual recognition proper to a moral and legal community of free and equal persons.

These are certainly evocative passages, suggesting that parents who genetically endow their children with characteristics they deem valuable unfairly handicap them in their struggle to achieve independence by planting a "fifth column" in their genome, which they cannot challenge, let alone overcome. But the appeal of this claim rests on a highly exaggerated contrast between genetic and social influence. Why should Habermas believe that it is any more hopeless to be at odds with the "genetically fixed" than the "environmentally fixed" intentions of a third person? To the extent that parents shape the character and abilities of their already-born children, they do so largely at a time when those children are too young to contest their influence in any coherent or effectual way, and they do so by "fixing" their intentions towards those children through such powerful mechanisms as habit-formation and internalization. These mechanisms may alter the child's psyche, and brain, as permanently as any genetic intervention.

A rearing parent always encounters an older child with the advantage of having had a profound influence on him at a time when the child could not effectively resist, whether shortly after conception or shortly after birth. A clever or recalcitrant child can reverse this advantage by blaming the parent, challenging her right to complain about features for which she bears responsibility, or absolving himself of responsibility for those features.

Moreover, as I noted earlier, very little genetic enhancement would be likely to bear fruit without the child's collaboration, or at least acquiescence. As skeptics about genetic technology are fond of pointing out, no conceivable genetic intervention, even the genetic replication of an actual person, could ensure that the resulting child would exploit his genetic endowment in the way his parents desired. Those parents would have to make the same intensive efforts as parents who did not avail themselves of genetic enhancement.

Habermas suggests no reason why a child should be less capable of reappraising values, habits, or skills promoted by genetic enhancement than those inculcated by early rearing. Almost all of his values, habits, or skills will reflect the interaction of the child's genome, engineered or not, with his rearing environment, and it is hard to see how an engineered child would be rendered incapable of reappraisal and resistance. To assume that genetic intervention cannot be resisted because it is too powerful is to embrace genetic determinism; to assume that it cannot be resisted because it is too integral to the child's identity is to embrace genetic essentialism.

Where Habermas appears to go astray is in emphasizing the indelibility of genetic interventions rather than their unilateral character. Parents who genetically enhance their future child would initially face only the vagaries of genetic manipulation, not the reflex resistance of an infant or toddler or the more self-conscious opposition of an older child or adolescent. The question is why this difference in the nature of the initial intervention should alter the moral landscape of the parent-child relationship.

Because of the unilateral role his parents initially play in shaping his attributes, a genetically engineered child may appear to be too much the creature of his parents. But more familiar ways of shaping one's children are also prone to excess, and the misgivings such excesses provoke may differ only in degree, not kind, from the misgivings we feel about genetic enhancement. Parents are often accused of treating their children as mere extensions of themselves, of failing to acknowledge and respect their status as separate persons, of disregarding or suppressing their distinctive natures by forcing them into a pre-set mold. Parents who genetically enhanced their future children would do something similar to overbearing parents, in failing to recognize, or show appropriate deference to, what a continental philosopher might call the "radical otherness" of the being with whom they are forming an intimate relationship. Their interventions are not consistent with the kind of restraint we require in relationships between moral equals, with the equality we expect parents to strive for long before it can be fully achieved.

The philosopher William Ruddick has suggested that parents have two distinct, and potentially incompatible roles: as gardeners, who cultivate desirable attributes, but also as guardians, who nurture the attributes their charges already possess. A successful parent balances the two roles, difficult as that is, cautiously shaping the nature of his children while carefully preserving the nature they already have. Genetic enhancement, like many more familiar rearing practices, might be seen as tipping the balance too far towards the gardening role, subordinating the child's nature, present or future, to the parents' own projects and ideals.

Control, Restraint, and Respect

The moral imbalance in unilaterally shaping a human being to one's own ideals may be illustrated by the original Pygmalion myth, though it concerns marriage rather than parenthood. Unlike his latter-day counterpart in Bernard Shaw's play, Pygmalion did not attempt to shape a living woman as a means to professional success or personal satisfaction. Resigned to a solitary existence, he had no intention of creating or remaking a human being:

> Yet fearing idleness, the nurse of ill,
> In sculpture exercised his happy skill;
> And carved in iv'ry such a maid, so fair,
> As Nature could not with his art compare. . . .

Yet he (like his latter-day counterpart) fell in love with what he created:

> Pleas'd with his idol, he commends, admires,
> Adores, and last the thing adored, desires.

He prayed to the Goddess of Love, Venus, and she made the statute come alive in his arms. Galatea, as he named her, reciprocated his love; they married, had a son, Paphos, and, apparently, lived happily ever after.

The original Pygmalion, then, was rewarded for his attempt to improve on nature with a passionate romance and a happy life. To us, however (or at least to me), there is something deeply troubling in his course of action—not in sculpting his ideal woman but in seeking to have her animated as his mate, after spurning all existing women. Were the Ancients simply more tolerant of such aspirations than we are? Or did Pygmalion really act unobjectionably, neither imposing his will and sensibilities on another human being, nor attempting to gratify himself by making another human being to his liking? After all, when he sculpted "Galatea" he was merely expressing what he thought was an impossible ideal, and by the time he prayed to have ivory become flesh, he was hopelessly in love. The qualities he sculpted into Galatea were not intended for his own benefit, save in the minimal sense of expressing his ideals. Once he saw those qualities embodied in ivory, it is not clear that he did anything wrong in seeking to have that matter brought to life.

But if so, what would be wrong in genetically "sculpting" a child to conform to one's ideals? Of course, unlike Pygmalion, the parents who employ genetic enhancement *would* expect their creation to come to life, and expect to enter into a long and intimate relationship with that creature. Moreover, the love of parents for a child is supposed to be even less conditional than the love of one spouse for another. But where is the fault, if parents attempt to shape their children to express their ideals as best they can, as opposed to catering to their vanity or material comfort, so long as they recognize, and accept, that those children will inevitably fall short of their ideals?

One response is theological. If children are a gift from God, they should be accepted as they are given. Ms. Manners has doubtless counseled many times that the recipient of a gift should try to avoid specifying what she wants, even if what she wants is noble—a contribution to Oxfam rather than a Lexus. But gifts come in many forms, which confer on the recipient widely varying degrees of control over what that recipient ultimately acquires. Why can't God confer a "gift certificate" good for a range of choices, rather than insist on making all the choices Himself? Clearly, such constraints on the kind of gift God can give have their source in a highly specific, far from universal, conception of the very modest role that human beings are assigned as God's co-creators.

Perhaps the theological recourse would be easier to resist if we had a richer secular conception of parenting and families; of what it means, or should mean, to help bring new people into existence and forge a singular kind of association to nurture them. As bioethicist Thomas

Murray has pointed out, our thinking about reproduction and families has been dominated by powerful but limited notions of negative liberty—liberty from government coercion and social pressure—at the expense of more positive notions of what it means to flourish as a parent, and in a family. Informed by "thicker" conceptions of the good of parenting and of families, we might be better able to articulate the wrongs or excesses of parental genetic enhancement. It may be that the kind of control sought by even the most conscientious genetic engineer is simply incompatible with the posture of openness and acceptance that parenthood and family require.

Many of us undoubtedly share the stubborn conviction that parents should accept the biological endowment their children receive by random genetic recombination. This conviction may draw its strength from the distinct one that we should accept our children "as they are." But that conviction has much more force when the alternative is rejection, not improvement. We certainly don't expect parents to accept what nature has dealt out to their *existing* children; we hardly discourage aggressive treatment for genetic diseases or accidental injuries, or many concerted efforts to improve a child's genetically constrained abilities. How can we object to attempts to improve on human nature prenatally without falling prey to genetic essentialism; without regarding any genetic improvement as if it were tantamount to rejection and replacement?

One promising approach, as I suggested earlier, is to place less emphasis on the genetic character of prenatal enhancement, and more on its unilateral character. On this approach, the importance of genes does not lie in their causal or constitutive role so much as in their independence from the will of their "donors." A child's existence as a distinct being begins with the randomness of genetic recombination. Such randomness may not be intrinsically valuable, but it serves the important function of limiting the control of parents and other agents. Genetic enhancement compromises that randomness and defies the limits it places on parental control.

On this approach, restraint in modifying genes is akin to restraint in modifying nature. To intervene without restraint is to fail to respect the independence or otherness of the beings and processes one encounters. While the analogy to nature obviously needs to be refined, it suggests that we can see in parental genetic enhancement the same kind of irreverence we see in uncontrolled development; and that we can condemn such irreverence in both domains without recourse to theology.

The analogy to nature also suggests, again, that an objection based on excessive control cannot be a categorical one; that we should regard the difference between prenatal enhancement and child rearing (and between enhancement and therapy) as a matter of degree rather than kind. We believe that parents should strive for a mean between control and acquiescence, and we can see genetic enhancement as falling toward the first extreme. But this leaves open the possibility that modest genetic enhancement will sometimes be less objectionable than ambitious rearing. Clearly, much work remains to be done to develop plausible secular objections to, and plausible moral constraints on, parental genetic enhancement.

I would like to thank Adrienne Asch, Alan Strudler, and Robert Wachbroit for their helpful comments during the development of this article.

Sources

Enhancing Human Traits: Ethical and Social Implications, edited by Erik Parens (Georgetown University Press, 2001); Alan Buchanan, Dan Brock, Norman Daniels, and Dan Wikler, *From Chance to Choice: Genetics and Justice* (Cambridge University Press, 2000); Nicholas Agar, "Liberal Eugenics," *Public Affairs Quarterly,* vol. 12 (1998); Leon Kass, "Babies by Means of *In-Vitro* Fertilization: Unethical Experiments on the Unborn," *New England Journal of Medicine,* vol. 285 (Nov. 18, 1971); W. Gardner, "Can Human Genetic Enhancement Be Prohibited?" *Journal of Medicine and Philosophy,* vol. 20 (1995); Paul Billings, Ruth Hubbard, and Stuart Newman, "Human Germline Gene Modification: A Dissent," *Lancet,* vol. 353 (May 29, 1999); Seana Shiffrin, "Wrongful Life, Procreative Responsibility, and the Significance Of Harm," *Legal Theory,* vol. 5 (1999); Dena Davis, "Genetic Dilemmas and the Child's Right to an Open Future, *Hastings Center Report* (March-April 1997); Matthew Clayton, "Individual Autonomy and Genetic Choice," in *A Companion to Genethics,* edited by Justine Burley and John Harris (Blackwell, 2002); Erik Parens and Adienne Asch, "The Disability Rights Critique of Prenatal Testing: Reflections and Recommendations," in *Prenatal Testing and Disability Rights,* edited by Parens and Asch (Georgetown University Pres, 1999); Glenn McGee, "Genetic Enhancement of Families," in *Pragmatic Bioethics,* edited by McGee (Vanderbilt University Press, 1999); Jürgen Habermas, "On the Way to Liberal Eugenics?" posted by the Colloquium in Law, Philosophy, and Political Theory (New York University) at: http://www.law.nyu.edu/clppt/program2001; William Ruddick, "Parents and Life Prospects in Having Children," in *Philosophical and Legal Reflections on Parenthood,* edited by Onora O'Neill and William Ruddick (Oxford University Press, 1979); Ovid, *Metamorphoses,* Book 10; Thomas Murray, "What Are Families For?: Getting to an Ethics of Reproductive Technology," *Hastings Center Report,* vol. 32 (2002).

The Ethics of Making the Body Beautiful: What Cosmetic Genetics Can Learn from Cosmetic Surgery

Sara Goering

Work to map the human genome is nearly complete, intensifying the debate about the appropriate uses of the information contained within this "book of life." We want to understand what these gene sequences make possible, and how they might be manipulated for good or for ill. We want to glean whether this knowledge will lead to new avenues for discrimination, or bridge such divides by highlighting the similarities in our biology. We ask ourselves whether we can avoid using our knowledge of the human genome for unethical ends.

Genetic manipulation for aesthetic reasons—cosmetic genetics—will be one of the important ethical challenges citizens must face in the future. The number of surgeries performed for cosmetic reasons has grown dramatically during this past decade, and it is plausible to believe that consumer demand will increase pressure to develop genetic techniques used for aesthetic enhancement. But we can recognize and debate those ethical challenges now, before techniques are developed which allow cosmetic genetics to become a part of an inevitable future reality.

Concerns about the ethics of cosmetic surgery offer important insights for cosmetic genetics. After briefly discussing what is meant by "plastic surgery," "cosmetic surgery," and "cosmetic genetics," this article explores one kind of argument commonly used in bioethics—the argument from precedent—to show that it cannot adequately discern or assess the ethical challenges posed by cosmetic genetics. The article then looks to some of the recent ethical attitudes toward cosmetic surgery in order to anticipate—and make recommendations about—the ethical challenges we will encounter when genetic therapies used for cosmetic purposes become a real option in the future.

The Popularity of Cosmetic Surgery

The term "plastic surgery" covers a broad range of surgeries that alter appearance. The term includes a wide range of reconstructive surgeries, which attempt to replace or repair congenitally malformed, damaged, or amputated areas of the body. Another subset of plastic surgery is cosmetic surgery, which is the topic of the present article. When used in this article, "cosmetic surgery" refers to surgery chosen primarily for aesthetic reasons or in hopes that one will become more socially acceptable. (In this discussion, "cosmetic surgery" does not refer to surgery intended to alleviate physical discomfort—as in breast reduction surgery, which relieves stress on the chest and back muscles caused by overlarge breast tissue—or which contributes to the physiological function of an individual.)

Insurance policies typically cover expenses incurred by reconstructive surgery, and some surgeries to correct functional disturbances (such as drooping eyelids that make seeing difficult). However, surgery for aesthetic reasons—cosmetic surgery—is widely available on a fee-for-service basis only. Despite its cost—a routine facelift is about $5,700—the popularity of cosmetic surgery is on the rise. According to the American Society of Plastic Surgeons, between 1992 and 1999 the number of cosmetic surgery procedures performed in the United States and Canada has risen 175 percent. Several types of surgery have seen an even more dramatic increase: liposuction has increased 389 percent and breast augmentation surgery has increased 413 percent.

Some anticipate a great market in genetic techniques applied for aesthetic enhancement. If one can choose surgery to create the bodily changes one desires, then why not choose genetic therapies to create those bodily changes or (by selective embryo implantation or the use of genetic therapies undertaken in vitro or during gestation) select desired physical traits for one's future children? In cosmetic genetics, the body *itself* produces such desired features as having blue eyes, being tall, maintaining a low body-fat ratio, developing larger breasts, or looking less "ethnic" (by designing nose shape, eyelid structure, hair texture, or skin color, among other features).

No genetic therapies exist today which make these options a reality, and many might be untroubled by the development of genetic techniques used for aesthetic enhancement, viewing cosmetic genetics as simply an extension of cosmetic surgery. But such a relaxed attitude would be a mistake, one which depends on accepting an "argument from precedent."

The Argument from Precedent

In anticipating the introduction of a new practice in medicine, citizens—and ethicists, too—commonly employ an "argument from precedent" to judge the ethical standing of the contemplated practice. That is, we compare the ends achieved by a new technology to those achieved by older accepted practices, and where these ends are similar, we conclude that the use of the new technology is morally permissible. Bioethics often relies on some version of the argument from precedent to assess the permissibility of new human genetic therapies. Since we treat genetic disorders such as cystic fibrosis to reduce their debilitating symptoms, so the argument goes, we ought to be willing to employ genetic interventions to eliminate diseases or treat their symptoms more effectively. Or, since we value childhood immunization, we ought to be willing to take advantage of genetic interventions to increase immunity. Applied in this way, the argument from precedent attempts to preserve morality by building on a foundation of previously accepted practices.

The argument from precedent is commonly appealed to, but rarely is it investigated fully, and recent work shows that its problems are significant. One difficulty is that it does not attend to morally relevant features in assessing the means used to achieve a desired end. Just because two different means accomplish the same general goal, we cannot assume they both achieve that goal in a moral way. We may share the goal of having our children learn well in school, for instance, but not find two alternate means—smaller class sizes versus increased use of Ritalin—ethically equivalent. Each strategy focuses on different "objects"—the child's environment, on the one hand, and the child's biology, on the other—and each leads to radically different experiences for the child. Decreased class size allows for more attention from teachers, more opportunity for students to express themselves, and a child's success relies on the *expression of his own* personality and talents. Increased prescription of Ritalin locates the problem *within* the child, suggesting to her that she is deficient and requiring that she *change herself* to meet the demands of others. Although it is not clear that the Ritalin option is necessarily immoral, it certainly deserves more sustained moral evaluation than it receives when we resort to the argument from precedent.

An even more basic, yet under-appreciated, problem with the argument from precedent is that it often does not include an independent ethical evaluation of commonly accepted practices. According to one understanding of the argument, concerns about cosmetic genetic enhancements for humans might be regarded as morally unfounded because cosmetic surgery is a popular and widely accepted method of altering one's physical appearance. Some might argue, in fact, that cosmetic genetics is *preferable* to cosmetic surgery because the techniques of cosmetic genetics eliminate the need for invasive surgery, which is an unavoidable part of many cosmetic procedures.

But this understanding obscures new ethical issues that arise with a new medical development. It also makes too quick a jump from what *is* practiced to what *ought* to be practiced. As Erik Parens, a bioethicist at the Hastings Center, notes, "There are many things we've always done that we think we ought not to do either now or in the future."

Further, cosmetic surgery is itself a hotly debated practice. Some critics have raised concerns over such issues as the quality of informed consent, the certification of plastic surgeons, and the riskiness of some procedures. But many feminist critics of cosmetic surgery emphasize deeper and more intractable moral issues, arguing that cosmetic surgery exacerbates "harmful conceptions of normality." These norms of appearance, they argue, are directed mainly at women, and specify what they *ought to look like* in a way that demands significant investments in time, energy, and money. Since most normal women cannot meet the societal ideal, even those with otherwise healthy, well-functioning bodies believe they have aesthetic "deficiencies" and feel dissatisfied with their corporeal lot. Feminist thinker Naomi Wolf says it well:

> When a modern woman is blessed with a body that can move, run, dance, play, and bring her to orgasm; with breasts free of cancer, a healthy uterus, a life twice as long as that of the average Victorian woman, long enough to let her express her character on her face; with enough to eat and a metabolism that protects her by laying down flesh

> where and when she needs it . . . the Age of Surgery undoes her immense good fortune. It breaks down into defective components the gift of her sentient, vital body and the individuality of her face, teaching her to experience her lifelong blessing as a lifelong curse.

A recent survey reports that 56 percent of women and 43 percent of men are dissatisfied with their overall appearance. Body doubles, air brushing, and digital magic help perfect the image of a societal ideal, and because many do not question the social pressure to achieve these unreasonable "norms," they contemplate—and many undertake—the risk of major surgery simply to approach that societal ideal. Although some of the procedures are fairly noninvasive and risk-free, others are painful, debilitating, and liable to cause permanent damage. Individuals recovering from facelifts can look and feel as though they have been seriously beaten, their payment of money—as well as swollen, reddened skin—in hopes of a long-term gain in aesthetic beauty.

The truth is, however, that those who undergo the surgery gain much *more* than just an aesthetic advantage. How one looks affects not only one's self-esteem and confidence, but also how others regard one's competence, personality, and likelihood for success. Even if the beauty standard is not fair or appropriate, from the perspective of rational self-interest it makes sense for individuals to undergo cosmetic surgery.

Yet if we think only of ourselves and the possibility for individual gain, we never contemplate the bigger picture and, when appropriate, act collectively. Because we want to think of ourselves as completely free agents, we deceive ourselves about our motivations and we become oblivious to the manipulation of others. With a narrow, individual focus, we may inadvertently act to sustain or reinforce harmful conceptions of normality rather than address their flawed assumptions. It is crucial to consider carefully why so many individuals currently pursue cosmetic surgery, how their individual actions shape the larger culture, and how their choices might spur developments in the even more tempting realm of cosmetic genetics.

Does Cosmetic Surgery Serve "Cultural Dopes"?

Although feminist thinkers generally agree that the pressures to conform to a youthful, slender, smooth-skinned, wide-eyed, often Eurocentric appearance are rooted in historical injustices, they disagree about how to understand the role of the individual in contributing to the popularity of cosmetic surgery. How one understands the relationship between the desires and motivations of the individual and the dictates of society leads to different strategies for addressing the problem of the pressure to conform to a "norm" of beauty.

One view of this relationship holds that women who undergo cosmetic surgery always do so wholly *because of* harmful norms, despite their claims to the contrary—they claim to be doing it for themselves. This view depicts women as passive "cultural dopes," controlled by their environment but unaware of that control. As feminist thinker Susan Bordo notes,

> People don't like to think that they are pawns of astute advertisers or even that they are responding to social norms. Women who have had or are contemplating cosmetic surgery consistently deny the influence of media images. "I'm doing it for me," they insist. But it's hard to account for most of their choices (breast enlargement and liposuction being the most frequently performed operations) outside the context of current cultural norms.

By participating in cosmetic surgery, these women flee from the realities of aging and change because traits associated with age are deemed unattractive by society. They want to avoid *being themselves,* but they claim to do it *for themselves.* In response to those women who claim to have finally discovered their real selves through cosmetic surgery (a claim that raises interesting issues of authenticity, akin to those patients of Peter Kramer who claim to have discovered their "real selves" through the use of Prozac), Bordo insists that such individuals both *deny* themselves the opportunity to understand our shared human condition of physical vulnerability, mortality, and impermanence, and they also *reinforce* harmful conceptions of normality through their actions. In effect, their actions increase pressure to fit the norm.

But if women who select cosmetic surgery are merely cultural dopes, then they seem to be absolved from responsibility for their actions. They simply follow the direction of outside forces that *shape* their desires. The best solution to the harmful conceptions of normality accepted by the "cultural dope" view is to change cultural pressures. This might be accomplished by demanding that the advertising industry present greater diversity in the body shapes of models. Careful regulation of the advertising industry might limit the creation of those new markets that rely on advertising aimed at *expanding* the scope of body image concerns. A more radical contingent might even find it appropriate to outlaw certain procedures. However, although the "cultural dope" view recognizes the myriad of strong cultural pressures exerting their influence on women, it denies that women are—or can be—free agents. Women are unthinking puppets of culture, and their behavior changes only because cultural norms change.

Does Cosmetic Surgery Create "Empowered Agents" (or Moral Hypocrites)?

Other feminist writers, such as Kathy Davis, argue that women who pursue cosmetic surgery are a picture of empowered agency. In her experience interviewing such women, Davis found that, rather than serving as "cultural dopes," these women were generally fully aware of the seemingly impossible system of appearance norms. Working as agents within their cultural constraints—yet cognizant of those constraints—they saw surgery as a "lamentable and problematic, but understandable course of action." In short, women choose the lesser of two evils: they act to attain the beauty norm rather than fall victim to it. Davis commends what she sees as women acting to control their identities. She reports that many women were "ashamed for feeling ashamed" of their bodies and chose cosmetic surgery *despite* strong objections from partners, friends, and family who offered constant reassurances about the women's natural beauty. Surprisingly, she found that even the women who did not have successful surgeries claimed that they had gained a better sense of their own agency and identity by their experience.

Although some good can come from adversity, it seems odd to commend a bad experience. Certainly one need not approve of the general situation that gives rise to it. In addition to her valorization of agency, Davis does not directly confront the fact that her interviewees appeared to hold one set of standards for themselves and another for other women. Each considered her own case exceptional, *she* had exceeded the "limit to how much suffering you should have to put up with" and suffered "more than what a woman should . . . have to endure." However, by Davis's own admission, most of these women were not obviously abnormal or atypical prior to their surgeries. Thus, the very consideration that Davis suggests makes these women more than cultural dopes—their ability to recognize the harmful norms that influence them and to make the best choices possible given these norms—seems to reveal hypocrisy (or at least some level of special pleading). By making exceptions in their *own* cases, these women illustrate their lack of commitment to their proclaimed general principle.

Moral evaluation of this situation cannot praise these women for their agency; instead, their choice raises questions of their integrity and the reasons for allowing personal exceptions.

Margaret Little offers another version of the "empowered agent" position. Suggesting that a change in beauty norms will take great effort, and probably could not be completed within one individual's lifetime, she argues that it would be an unjustifiable sacrifice to deny cosmetic surgery to individuals who suffer *today* because of their bodily condition. Little concludes that it would be morally permissible for surgeons to continue to provide cosmetic surgery so long as they work at the same time to change the very norms that bring them most of their customers:

> If one must perform surgeries to help people meet suspect norms of appearance (out of concern for their suffering, say) then one must maintain an overall stance of fighting the norms. The only way to participate in the surgeries without de facto promoting the evil whose effects one decries is to locate the surgery in a broader context of naming and rejecting the evil norms. One's purpose and meaning—that of alleviating the extreme burdens the system places on some—can be expressed only if one's broader actions stand squarely against the norms.

By "broader actions" Little means that cosmetic surgeons should "speak out against the suspect content of the norms" both in public and in their private consultations with patients. Cosmetic surgeons ought to discuss with prospective patients the option of not having any surgery at all, and they must clarify the risks and possible side effects of contemplated procedures.

However, it *already* is common practice for cosmetic surgeons to assess the surgical and "emotional" success of procedures their patients contemplate. It is also routine to discuss with patients their expectations, and to inform them of risks and other options available to them. Even if cosmetic surgeons did not do what Little advocates, her suggestion seems strange because it relies on the very person who benefits from the women's desire for surgery (both financially and psychologically, since surgeons derive personal satisfaction from their skill) to try to *eliminate that desire.* Placing the responsibility for revising the norm in such hands is likely to create minor change, if any. Little might also ask women who undergo cosmetic surgeries to speak out against the harmful norms that influenced their decisions. Surely this would be even stranger. Most women hesitate to discuss their surgeries, and those who do would find themselves in the odd position of telling others not to do something that has made them individually better off. One can hardly expect a surgically altered, societally perfect advocate for changing beauty standards to be taken seriously. Adopting this tactic avoids sacrificing women to social change only to limit their capacity to promote social change.

Can Cosmetic Surgery Contribute to the "Revalorization of the Ugly"?

Is there any way to *recognize* the suspect norms, *accept* the practice of cosmetic surgery, and *avoid* the conclusion that women who receive it are either cultural dopes or apparent hypocrites? Kathryn Morgan proposes a fairly shocking response to this problem. She suggests that women ought to "take back" cosmetic surgery and use it to highlight the arbitrariness of the cultural norms that currently lead women to choose cosmetic surgery. In order to "revalorize the ugly" Morgan proposes (tongue-in-cheek) that women start requesting skin wrinkling procedures, fat injections for their thighs, and techniques specifically designed to make their breasts and eyelids sag. Her proposal intends to show both the strength and the arbitrariness of the current beauty norms. If we are horrified to think of women undergoing drastic and unnecessary surgical measures to make a point, then we should also be horrified to think of women undergoing drastic and unnecessary surgical procedures to gain social acceptability.

French performance artist Orlan might be a case for Morgan, although Orlan's nine cosmetic surgeries have been aimed more at critiquing the possibility of the ideal body than at specifically creating ugliness. Orlan has attempted to make her face resemble a compilation of the facial structures of beautiful women painted by great artists, in order to "show, by example, that the legacy of masculine portrayals of feminine beauty precludes women's full agency and control." To this end, she has had, for example, silicone implanted in her forehead to make it more closely resemble the forehead of Mona Lisa. Her pursuit of cosmetic surgery is a political act. She is "not against *all* cosmetic surgery, but against *the way it is used*"—to make women fit a code of feminine beauty that requires conformity rather than individuality.

Lessons for Cosmetic Genetics

Several lessons can be learned from this brief survey of the ethics of cosmetic surgery. One learns that when suspect social norms are at the *root* of a practice and are themselves *reinforced* by continued patronage of it, one at best achieves only temporary and personal comfort by continuing the practice. Davis admires the protagonist of Fay Weldon's novel *The Life and Loves of a She-Devil*, for:

> She does not see cosmetic surgery as the perfect solution and she is well aware of the enormous price for women who undertake it. Under the circumstances, however, it is the best she can do. For she knows only too well that the context of structured gender inequality makes this solution—as perhaps any solution—at best, a temporary one.

However, in acting for individual comfort, one undercuts larger societal goals. Further, societal norms at times seem intractable only because they require collective action for change.

The debates about the ethics of cosmetic surgery can inform the coming debate over the appropriateness of cosmetic genetics. But even before cosmetic genetics becomes a reality, citizens can recognize its dangers and take action to enact legislative bans, distribute research funds in a thoughtful way, and initiate widespread public education programs. Prudence suggests placing a temporary moratorium on public funding for genetic research designed to identify or offer therapy to alter primarily cosmetic traits. Certainly, devastating genetic disorders must have priority.

If cosmetic screening tests or genetic therapies eventually become available (through private or corporate research, or through extensions of approved federally funded research), hospitals and clinics should impose regulations that restrict the use of such tests. Expectant parents often want as much information as possible about their future child, but clinics can determine when such tests are appropriate, or refuse to employ them altogether.

Finally, one cannot overemphasize the need for a broad public education program. Even if hospitals and clinics impose their own restrictions, it seems likely that entrepreneurs will step forward eagerly to offer such services outside the regular medical setting. The best way to combat that issue is to address market demand. Public education programs that emphasize health, and promote the beauty and uniqueness of diverse body shapes, would help all of us be more satisfied with our bodies (and more likely to accept a future child who does not fit the ideal). With sincere effort, we might be able to abandon an ideal based on a specific physical body type and embrace an ideal that emphasizes such deeper commitments as participation in society, intellectual prowess, and emotional caregiving. Better funding for programs that focus on these deeper commitments might accelerate change. For instance, Girls Incorporated is a national program that aims to help young girls "confront subtle societal messages about their value and potential." Included in the program is a Bill of Rights that stresses the "right to accept and enjoy the bodies [girls] were born with and not to feel pressured to compromise their health in order to satisfy the dictates of an 'ideal' physical image."

Cosmetic genetics can learn this lesson from cosmetic surgery: if a practice contributes to or reinforces harmful conceptions of normality, we should look for other means to achieve individual interests. We often dismiss alternatives too quickly because we cannot be certain that other people will follow suit, and if they do not, we might put ourselves at a disadvantage. But social change does not happen on its own. The answer is one that promotes agency, but not agency with moral blindfolds. No doubt we ought to respect individual choices, and to support individuals who feel unduly pressured. At the same time, however, we must be willing to criticize the choices that stem from individual agency, especially when those choices ignore the harmful conceptions of normality or unfairly create special exceptions for individuals. We certainly cannot benefit our children by making them the "perfect" offspring of cultural dopes or moral hypocrites.

Sources

For facts and statistics concerning cosmetic surgery, see: Kathy Davis, *Reshaping the Female Body: The Dilemma of Cosmetic Surgery* (Routledge, 1995), http://www.plastic-surgery.net/fees.html, www.plasticsurgery.org/mediactr.htm, and http://surgery.org/media/statistics/quickfacts1.html. Two works relying on an argument from precedent are: Glen McGee, *The Perfect Baby* (Rowman and Littlefield, 1997) and John Harris, *Wonderwoman and Superman: The Ethics of Human Biotechnology* (Oxford University Press, 1992); Erik Parens, "Is Better Always Good? The Enhancement Project" in *Enhancing Human Traits: Ethical and Social Implications,* edited by Erik Parens, Mark J. Hanson, and Daniel Callahan (Georgetown University Press, 1998); also, Ronald Cole-Turner, "Do Means Matter?" and Dan Brock, "Enhancements of Human Functions: Some Distinctions for Policymakers" are found in the Parens et al. volume. Ethical issues of Ritalin use is discussed by Claudia Mills, "One Pill Makes You Smarter: An Ethical Appraisal of the Rise of Ritalin," *Report from the Institute of Philosophy & Public Policy,* vol. 18 (1998); the phrase "harmful conceptions of normality" is used by Margaret Little, "Cosmetic Surgery, Suspect Norms, and the Ethics of Complicity" in Parens et al. volume; Naomi Wolf, *The Beauty Myth: How Images of Beauty Are Used Against Women* (Anchor Books/Doubleday,

1991); a *Psychology Today* survey reports that 56 percent of women and 43 percent of men are dissatisfied with their overall appearance (Jan./Feb. issue 1997); Susan Bordo, "Braveheart, Babe, and the Contemporary Body," in Parens et al. volume. For discussions concerning Philosophy & Public Policy Quarterly the possibility of finding one's "real self" through medical interventions, see: Peter Kramer, *Listening to Prozac* (Penguin Books, 1997) and the collection of articles in the *Hastings Center Report*, vol. 30 (2000). Recent examples of responses to marketing pressure are Britain's Body Image Summit (June 2000) and the Real Women Project, a southern California effort, discussed in "Real Women Take on a Real Image Problem," *Los Angeles Times* (July 5, 2000) and http://www.realwomenproject.com; Kathy Davis, *Reshaping the Female Body* (Routledge, 1995) and also her "The Rhetoric of Cosmetic Surgery: Luxury or Welfare?" in the Parens et al. volume; Margaret Little, "Cosmetic Surgery, Suspect Norms, and the Ethics of Complicity," in the Parens et al. volume. For a discussion of surgical and "emotional" success, see Thomas Pruzinsky, "Cosmetic Plastic Surgery and Body Image: Critical Factors in Patient Assessment" in *Body Image, Eating Disorders and Obesity,* edited by J. Kevin Thompson (American Psychological Association, 1996). On the hesitancy of women to discuss their surgeries, see Susan Zimmerman, *Silicone Survivors: Women's Experiences with Breast Implants* (Temple University Press, 1998); Kathryn Pauly Morgan, "Women and the Knife: Cosmetic Surgery and the Colonization of Women's Bodies" in *Sex/Machine*, edited by Patrick Hopkins (Indiana University Press, 1998); the performance artist Orlan aimed at gaining "the chin of Sandro Botticelli's *Birth of Venus*, the forehead of da Vinci's *Mona Lisa*, the lips of Gustave Moreau's *Abduction of Europa*, the eyes of a Fountainebleu School *Diane Chasseresse*, and the nose of Gerard's *First Kiss of Eros and Psyche*," according to Peg Brand, "Bound to Beauty: An Interview with Orlan" in *Beauty Matters*, edited by Peg Brand (Indiana University Press, 2000). The issue of genetic therapies for genetic disorders is a complex one. The disability rights movement, for instance, notes that many conditions the general public would be tempted to "cure" through genetic therapies may not in fact rightly be considered physiologically undesirable. For a discussion of this debate, see *Prenatal Screening and Disability Rights*, edited by Erik Parens and Adrienne Asch (Georgetown University Press, 2000), and Anita Silvers, David Wasserman, and Mary B. Mahowald, *Disability, Difference, Discrimination: Perspectives on Justice in Bioethics and Public Policy* (Rowman and Littlefield, 1998); on limits to cosmetic screening and genetic therapies, see Jeffrey Botkin, "Developing Professional Standards for Diagnostic Services" in the *Prenatal Testing* volume and also Parens's "Is Better Always Good? The Enhancement Project" in the Parens et al. volume. For information on Girls Incorporated, see its web site at http://www.feminist.com/girlsinc.htm.

Index

About the Editor and Contributors

Harold W. Baillie is professor of philosophy at the University of Scranton, and he is also on the faculties of the University of Trnava in Slovakia and the State Medical University in Tbilisi, Republic of Georgia. He has written articles on subjects ranging from health care ethics to metaphysics and political philosophy, and is coauthor of *Health Care Ethics: Principles and Problems* (with Drs. Thomas and Rosellen Garrett), which is in its fourth edition from Prentice-Hall. He is currently working with the ethics committees of several local and regional health care facilities, and he serves as the chief compliance officer for the Community Medical Center Health System.

William A. Galston is director of the Institute for Philosophy and Public Policy, Saul I. Stern Professor of Civic Engagement at the School of Public Affairs at the University of Maryland, and director of the Center for Information & Research on Civic Learning & Engagement (CIRCLE). He is a political theorist who both studies and participates in American politics and domestic policy. He was deputy assistant to the president for Domestic Policy, 1993–1995, and executive director of the National Commission on Civic Renewal, 1996–1999. Galston served as a founding member of the Board of the National Campaign to Prevent Teen Pregnancy and as chair of the campaign's task force on religion and public values. He is the author of five books and nearly one hundred articles in moral and political theory, American politics, and public policy. His publications include *Liberal Purposes* (Cambridge, 1991) and *Liberal Pluralism* (Cambridge, 2002).

Verna V. Gehring is editor at the Institute for Philosophy and Public Policy at the School of Public Affairs, University of Maryland. She is a philosopher broadly interested in the obligations of state and citizen, and the various accounts of civil society. In addition to her work on the seventeenth-century political philosopher Thomas Hobbes and his enduring influence, Gehring's interest is applied to such contemporary matters as the state lottery, nuclear proliferation, computer hackers, and the social harms caused by imposters. She is editor of *Philosophy & Public Policy Quarterly,* coeditor (with William A. Galston) of *Philosophical Dimensions of Public Policy* (2002), and editor of *War after September 11* (2002).

Sara Goering is assistant professor of philosophy and director of the Center for Applied Ethics at California State University, Long Beach. Her work combines concerns about the ethics of genetic engineering, prenatal testing, human enhancement, disability rights, and feminist theory. A central theme is the definition and significance of normality. She is coeditor, with Annette Dula, of *"It Just Ain't Fair": The Ethics of Health Care for African Americans* (Praeger Publishers, 1994). She is currently working on an interdisciplinary survey study of cosmetic surgery, aging, and norms of appearance.

Deborah Hellman is associate professor at the University of Maryland School of Law. Her research interests include constitutional law, theories of discrimination, and the ethics of clinical medical research. Her articles include, among others, "The Expressive Dimension of Equal Protection," *Minnesota Law Review* (2000); "Two Types of Discrimination: The Familiar and the Forgotten," *California Law Review* (1998); "Is Actuarially Fair Insurance Pricing Actually Fair?: A Case Study in Insuring Battered Women," *Harvard Civil Rights-Civil Liberties Law Review* (1997); "The Importance of Appearing Principled," *Arizona Law Review* (1995); and (with Samuel Hellman) "Of Mice But Not Men: Problems of the Randomized Clinical Trial," *New England Journal of Medicine* (1991).

Mark Sagoff is research scholar at the Institute for Philosophy and Public Policy at the School of Public Affairs, University of Maryland. He has been a Pew Scholar in Conservation and the Environment, past president of the International Society of Environmental Ethics, and a Fellow of the American Association for the Advancement of Science. He has published widely in journals of philosophy, law, economics, and public policy. His book *The Economy of the Earth: Philosophy, Law, and the Environment* (1988) has received wide acclaim.

Paul B. Thompson became the W. K. Kellogg Endowed Chair in Food, Agricultural, and Community Ethics at Michigan State University in August of 2003, after teaching philosophy at Purdue University and Texas A&M University for a combined total of over twenty years. His research has addressed all manner of controversial issues in the US and global food systems. He has served as a consultant to the Food and Agricultural Organization of the United Nations and he is a member of the National Research Council's Advisory Committee on Agricultural Biotechnology. He is the author of numerous articles, and his books include *The Spirit of the Soil: Agriculture and Environmental Ethics* (1995) and *The Agrarian Roots of Pragmatism* (edited with Thomas C. Hilde, 2000).

Robert Wachbroit is research scholar at the Institute for Philosophy and Public Policy at the School of Public Affairs, University of Maryland. He is also adjunct associate professor of OB/GYN in the University's School of Medicine. He has written articles in the areas of philosophy of science, philosophy of medicine, and medical ethics, including articles on the principles of disease classification, the challenges of genetic testing and diagnosis, and the problem inherent in human cloning and genetic enhancements. He has also written about the role of expertise in public deliberations and on the impact of the Internet on civil society. He is coeditor (with David Wasserman) of *Genetics and Criminal Behavior* (Cambridge, 2001) and (with David Wasserman and Jerome Bickenbach) *Quality of Life and Human Difference: Genetic Testing, Health-Care, and Disability* (Cambridge, forthcoming).

David T. Wasserman is research scholar at the Institute for Philosophy and Public Policy at the School of Public Affairs, University of Maryland, and a practicing attorney. His work focuses on ethical and policy issues in genetic research, reproductive technology, health care, and disability. In addition to many articles and book chapters, he is coauthor of *Disability, Difference, Discrimination,* with Anita Silvers and Mary Mahowald (1998) and coeditor (with Robert Wachbroit) of *Genetics and Criminal Behavior* (Cambridge, 2001) and (with Robert Wachbroit and Jerome Bickenbach) *Quality of Life and Human Difference: Genetic Testing, Health-Care, and Disability* (Cambridge, forthcoming).

Richard M. Zaner recently retired as Ann Geddes Stahlman Professor Emeritus of Medical Ethics and Philosophy of Medicine, Vanderbilt University School of Medicine. Founder and director of the Center for Clinical and Research Ethics, he served as clinical ethicist and director of the Clinical Ethics Consultation Service for Vanderbilt Hospitals, and founded and cochaired the Medical Center Ethics Committee. He held secondary appointments as professor, Department of Philosophy; professor of ethics, Graduate Department of Religion; and professor of ethics, Divinity School; adjunct professor of health care ethics, School of Nursing; and research member, with scholarly status, Kennedy Center for Research in Education and Human Development, where he cofounded the Ethics Service for Research. His scholarly interests are mainly concerned with developing the phenomenological approach to (individual and social) human life. Zaner has authored six works; his collection of clinical ethics narratives, *Troubled Voices,* was selected "Outstanding Academic Books of 1994" by *Choice.* He has edited eleven professional books, translated three books and a number of essays, and published more than 125 sections of books and professional journal articles.